Philippe P. THEYS
Graduate Engineer from École Centrale de Paris

QUEST FOR QUALITY DATA

Foreword by
C. Lwanga Yongke
AeraEnergy

2011

 Editions TECHNIP 25 rue Ginoux, 75015 PARIS, FRANCE

FROM THE SAME PUBLISHER

- Log Data Acquisition and Quality Control. Second Edition
  P. THEYS
- Heavy Crude Oil
  From Geology to Upgrading. An Overview
  A.Y. HUC
- CO_2 Capture
  Technologies to Reduce Greenhouse Gas Emissions
  J. LECOMTE, P. BROUTIN, E. LEBAS
- Multiphase Production
  Pipeline Transport, Pumping and Metering
  J. FALCIMAIGNE, S. DECARRE
- Corrosion and Degradation of Metallic Materials
  Understanding of the Phenomena and Applications in Petrolem and Process Industries
  F. ROPITAL
- A Geoscientist's Guide to Petrophysics
  B. ZINSZNER, F.M. PERRIN
- Acido-Basic Catalysis (2 vols.)
  Application to Refining and Petrochemistry
  C. MARCILLY
- Petroleum Microbiology (2 vols.)
  J.P. VANDECASTEELE
- Physico-Chemical Analysis of Industrial Catalysts
  A Practical Guide to Characterisation
  J. LYNCH
- Petrochemical Processes (2 vols.)
  Technical and Economic Characteristics
  A. CHAUVEL, G. LEFEBVRE
- Well Logging Handbook
  O. SERRA
- Well Logging and Reservoir Evaluation
  O. SERRA

Printed in France

ISBN 978-2-7108-0964-7

to the field engineers…

Acknowledgements

Substantial help was received during the completion of this book.

Peter Ireland came out with the idea of using bathroom scales to better clarify the difference between reality and measurements. Peter Fitzgerald elucidated a number of details related to depth issues.

A few members of the Oilfield Data Quality discussion group on the Linkedin website accepted to review the draft of the book. Thal Mc Ginness was particularly active and digested four chapters. Other reviewers were Tom Barber, Claude Baudouin, Jean-Alain Chodorge, Robert Cluff, Charles Flaum, Mauro Gonfalini, Guy Goyeau, Shujie Liu, Gary Myers, Annie Roulet and Alan Sibbit. The review is theirs, but the remaining mistakes are mine. Mark Hutchinson almost entirely wrote Chapter 18 on drilling data.

The oil companies, AeraEnergy, bp, Chevron, ConocoPhillips, ETAP, ExxonMobil, Marathon, Occidental, Petroleum Development Oman and Total opened their archives in the last seven years. These logs definitely contributed to an increased understanding of the issues on data quality.

Finally, Editions Technip are to be thanked for the support they provided in the publication of a text with numerous technical difficulties.

Foreword

We have heard the now familiar statement. Reserves are the primary asset of an oil company. And these reserves are based on data. So perhaps the real asset of an oil company is the data it has about these reserves.

If we truly believed this, as an oil industry, would we take better care of our well log data?

My personal quest for quality well log data began in earnest in 1999, as I led a team of dedicated professionals through a two-year data migration project. At the conclusion of the project, the team had loaded over 45,000 wells and 1.2 million log curves in the company's Geoscience database. The heart of the project was the extraction, scrubbing, consolidation and migration of the database. As the project neared completion, I remember being driven by two thoughts: a) how do I make sure we never have to do this again and 2) how do we even know if all the measurements these logs represent are accurate?

As my organization was setting out to drill, log and complete close to 1,000 wells per year, for several years, these were not trivial concerns.

Enter Philippe Theys. From his first book, *Log Data Acquisition and Quality Control*, from his courses and through numerous conversations over the past 11 years, we determined how to apply to the specific domain of well logs, data quality principles we had learned and implemented for other subject areas. Philippe provided the science, the techniques and the insights we needed to assure the logs we acquired were as complete and accurate as possible.

In *Quest for Quality Data*, Philippe continues to dispel the mysteries that typically surround well measurements and their uncertainties. He also illustrates how to incorporate into the well logging process two of the fundamental principles of data quality management:

- Assure and manage data quality at the source,
- Manage data not as a by-product, but as the carefully planned product of a well-defined "data production process", involving data suppliers (the logging companies), and data consumers, the oil company's petrophysical engineers and geoscientists.

Philippe's call for a new and higher sense of ethics, discipline and technical excellence in the well logging discipline stands as a manifesto for the profession.

As Philippe writes in this book, "Formation evaluation has two ingredients, measurements and models." So far, models and information technology have garnered most of the attention. The current focus of the digital oilfield implementations that are sweeping the oil industry also shows this disturbing trend.

By shifting the spotlight to data and information, pioneers such as Philippe are ensuring that the Information Age does not become reduced to a mere Information Technology Age. Data and information must indeed be valued and managed as the currency of the knowledge economy.

This is an exciting time for me. After working in information quality for more than a decade, I am starting to see tangible evidence that information quality is finally establishing itself as a real profession, with a rigorous academic foundation, a wide body of knowledge, distinctive best practices nurtured and honed by practitioners in the front line. The signs are many and include the following:

- A professional society, the International Association for Information and Data Quality (IAIDQ), is celebrating its sixth anniversary in 2010.
- The upcoming Information Quality Professional certification developed by IAIDQ.
- The publication of several ISO standards dealing with information quality: ISO 20512 and ISO 8000.
- Information and data quality conferences now being held around the world.
- More job listings for information/data quality professionals, and the increasing maturity of the work expectations: there is now a broad understanding that information/data quality is not and has never been synonym to data cleansing! A growing number of job postings now require process improvement experience and skill in Six Sigma and Lean.
- Information and data quality courses and degrees offered by colleges and universities around the world, led by the University of Arkansas at Little Rock which boasts the first Masters and PhD degrees in information quality in the United States and perhaps the world.
- Books such as Philippe's *Quest for Quality Data* and *Log Data Acquisition and Quality Control.*

If quality information is the primary input to reservoir management analysis and decision-making and analysis, then the measurement data collected down the wellbore is the vital raw material of this value chain. Oil companies wise enough to apply the prescriptions that Philippe lays out in this book will take giant forward strides in their ability to properly manage their most important data asset. As a result, they will dramatically enhance their ability to capture the voice of the reservoir and to hear the tale of the well, in order to maximize profitable reserve recovery.

C. Lwanga Yongke
Information Quality Process Manager, Aera Energy LLC
Founding Member and Advisor to the Board of Directors, IAIDQ

Contents

Chapitre 3

ALL WELL MEASUREMENTS ARE INDIRECT

Chapitre 4

LOGGING MEASUREMENTS DO NOT FOCUS ON ZONES OF INTEREST

Chapitre 5

MEASUREMENTS ARE IMPRECISE AND INACCURATE

Chapitre 6

HOW MEASUREMENTS CAN SUFFER FROM HUMAN BIAS

Chapitre 7
COMPLEXITY

Chapitre 8
COMPLICATION

Chapitre 9

WYSINWYTII

Chapitre 10
MISCONCEPTIONS

Part 2
QUEST FOR QUALITY DATA

Chapitre 11
THE DIFFERENT USES OF LOGGING DATA

Chapitre 12

BROCHURE SPECIFICATIONS

Chapitre 13

QUEST FOR UNCERTAINTIES: FROM BROCHURE SPECIFICATIONS TO REAL UNCERTAINTIES

Chapitre 14
DELIVERABLES

Chapitre 15
DEPTH

Chapitre 16
HIDDEN TREASURES

Chapitre 17

CONTRIBUTION OF THE FIELD ENGINEER TO THE QUALITY OF DATA

Chapitre 18

DRILLING DATA

Chapitre 19

CORING DATA

Chapitre 20

CONCLUSIONS AND RECOMMENDATIONS

APPENDIX

Preface

Why this book? The first edition of *Log data acquisition and quality control* has been assembled more than twenty years ago. Since, the technology used by the oil industry has considerably changed. Still the fundamental issues of database and uncertainty management are very much alive. In the mean time, my work has widened to include a broader perspective and understanding of the oil companies, especially through consulting for them. A book highlighting the challenges of the industry in its quest for quality data that enables effective decision-making and quantitative analysis is needed.

My approach to data quality has been marked by three epiphanies:

The first one took place early in my career, in the mid 70s, when I was a field engineer collecting data in an exploration well. I discovered that my client, an oil company, was more interested with being delivered with what they were expected than with real, objective information. It took them weeks to finally accept that they were facing a duster, a fact that was clear from the logs run early on. I had done a good job, but they were rather upset about my work. Messengers are not popular when they bring bad news.

The second epiphany occurred after my assignment as the head of interpretation studies in the early 80s. My section helped in developing the interpretation processes for the measurements derived from the LDT*[1] (Litho-density), NGT* (Natural Gamma Ray Spectroscopy), DRI[2] (Dual Resistivity), DPT (Deep Electromagnetic Propagation), EPT (Electromagnetic Propagation) and SHDT (Stratigraphic High-resolution Dipmeter) tools, a generation of equipment that attempted to bring quantitative information. Gone back to the field organization, I found:

- that the field processes were not rigorous enough to match the needs of our interpretation programs,
- that the drilling conditions were dramatically hurting the production of useable data.

I spent a great deal of time to make the field organization aware of the enhanced process requirements and the oil companies knowledgeable about the effect of drilling on data acquisition. I was encouraged by my contacts with the drilling departments of large oil companies. They can be summarized by the statement of a drilling manager: "I had no idea that

1. An asterisk denotes a mark of Schlumberger.
2. This tool was not commercialized.

we were doing so much harm to data and information. We'll cooperate with you." He developed a set of recommendations that were quickly published through his organization.

The third revelation happened upon my retirement, at the beginning of my career as a consultant. I worked for a dozen of oil companies. I discovered that the use of the data was sometimes far from the one I worked on as an employee of a logging company.

From these successive understandings, I gathered that there is a tremendous amount of miscommunication in the industry. I summarize the main points, which will be further investigated in the rest of the book:

- Logging companies acquire measurements, not real values.
- Well logging companies are not service companies. They are product companies.
- Data is acquired for multiple uses, not only for quick, quasi-real time decisions, but also for the long-term. The quality requirements for these various usages are quite different.
- What one sees on a log or in a digital log data base is sometimes not what one thinks.

The field engineer

This book is dedicated to the field engineers who work in the oil industry. They have a very challenging and important mission. They provide the first link in the data acquisition and interpretation process.

The perfect log

The previous paragraphs may sound negative, but there is hope. It can be represented by the following anecdote. My colleagues and friends in oil companies and data acquisition companies often joke about my search for the perfect log. It is true that after reviewing several thousand logs, I have only found a number of excellent logs that could be counted with the fingers of two hands. A few years ago, an oil company representative invited me to join him in a meeting room. There were a dozen people in the room. On the table, in the middle of the room was the thick print of a recently acquired log. I was required to look at it. After twenty minutes spent scrutinizing the document in silence, I summarized my finding: "This looks like a very good log, complete, neat and well documented." The oil company person replied: "Yes, for us, it is a perfect log. **It can be done**!"

Philippe Theys
Houston, 2010

Three notes

In order not to complicate the grammatical structure of some paragraphs, I have used the masculine "he" or "his." This is by no mean a form of disrespect to the growing number of women in the industry. "She" or "her" could have been used throughout the book.

Data is indeed the plural of datum, a Latine word. The modern American English usage, data as a singular, just as in sugar, is my choice.

Proper abbreviations from the International System of Units (SI from the French Le Système International d'Unités) are used. So, the reader will find "s" instead of "seconds" and "h" instead of hours, etc.

1

Introduction

This book is not another edition of *Log Data Acquisition and Quality Control*. And it is not a training book on the most recent logging tools or interpretation techniques. It is still, though, concentrating on the acquisition of data.

It analyzes first the misconceptions that prevail about well data and then brings some recommendations on how to guarantee the proper collection of data that will elegantly go through the brunt of time.

In Part 1, the reader learns about the hurdles that prevent an optimal use of the data. The first major difficulty relates to the fact that well data does not describe reality, but measures it. Simple examples of measurements are presented. It is then explained that almost no well measurement is direct and that the full process has many intermediate steps. The issue of volume of investigation, never in coincidence with the piece of rock the data user wants to know about is also raised. Basic truths about measurements, always handicapped by systematic and random errors, are exposed in the following chapter.

Human bias is then considered. Nudging, fudging and shifting do happen. The name of "services companies" attributed to the data vendors, is part of the problem. The reader is then exposed to the incredible changes of the industry. The logging company is submitted to a dramatic complexity, horizontal and sinuous wells, exotic muds and remoter well sites. The data user, on his side, faces complicated logging equipment and data sets. Presentations are obscure and sometimes ambiguous.

The reader may be struck by pessimism by this dark diagnostic. In Part 2, most problems are accompanied by a solution. First, different data usages call for different acquisition approaches. It is also explained how the cost of data is offset by multiple and successive use of the data sets.

Well logging data, regardless of conveyance is used as an example in most of the chapters. Chapter 18 extends the application to drilling data. Chapter 19 indicates that cores' data cannot be the ultimate reference and is also affected by a number of shortcomings.

Part 1

Why measurements differ from reality

2

Setting the problem with simple examples

2.1 MEASUREMENTS AND REALITY

Without getting into lengthy philosophical discussions, it is to be stressed that a human being cannot embrace reality. Reality is unknowable. Senses allow a description of reality that is filtered. For instance, the human ear does not perceive low acoustic frequencies below, say, 20 Hz, and high frequencies above 16,000 Hz. The human eye does not see past infrared and ultraviolet. The perception of reality is therefore distorted.

A quantitative assessment of reality is performed with measurements. Four quantities intervene often in physics: length, mass, time and current intensity. From them, many more attributes of the physical world can be derived: velocity, acceleration, strength, etc. Even the measurements of the basic four quantities are difficult. Most people wear a watch today, but it is only two hundred years ago that time pieces became accurate and precise. Mechanical chronometers, quartz watches and atomic clocks have exceedingly complex technologies even though it is available. Current intensity is measured with a multi-meter device, which contains dozens of electric components.

It is necessary to understand the harsh truth on measurements before using any of them, and this includes measurements performed in wells. Of help is metrology, the science of measurement, developed two centuries ago. Metrologists are the experts of this science. To understand the challenges encountered while performing measurements, two common and relatively simple measurements, measuring the length of an object and weighting a human body with a bathroom scale, are considered. The two measurements are combined to yield BMI, the body mass index.

2.2 MEASURING THE LENGTH OF A STICK

The objective is to measure the length L of a stick. The measurement of length is one of the simplest processes in physics. Two rulers are used, one graduated in cm, the other one in inches. Fig. 2.1 displays what the observer can see, blown up about three times. With a normal eyesight, he will conclude:

- with the cm ruler: 29.4 cm < L < 29.5 cm
- with the in ruler: 11 9/16 in. < L < 11 10/16 in.

With the cm-ruler, the relative error would be of the order of 30 to 90 ppm. [1] With the inch-ruler, it would be 0.2 to 0.4%. For most purposes, these measurements can be qualified as excellent, though laser micrometers can achieve a 5-µm repeatability (about 1 ppm). In comparison, the accuracy of logging measurements is seldom better than a few %.

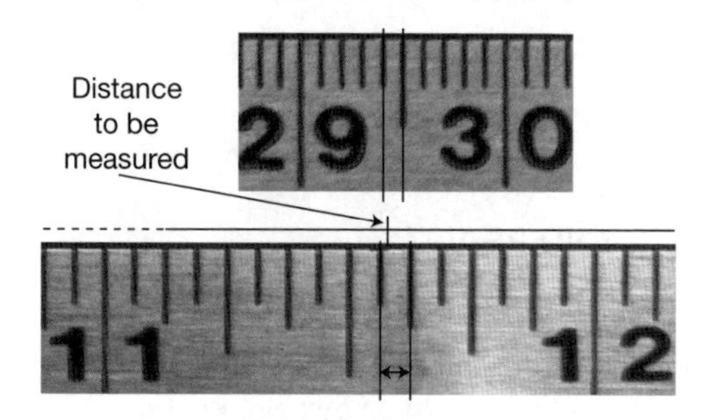

Figure 2.1

Measurement of a stick with two rulers (graduated in cm, top, and in, bottom). The cm-ruler is shown as an inverted image to facilitate the comparison with the in-ruler.

Measurement of length in the example is simple. Nevertheless, it yields a range of values and not a single value. This is the case for any measurement.

2.3 MEASURING THE WEIGHT OF A PERSON WITH A BATHROOM SCALE

Mass [2] is monitored in many engineering projects. Its measurement is considered in a simple and common project, the measurement of the weight of a human being.

A person has for objective to lose weight. To monitor progress, it is necessary to perform successive measurements of weight, starting from a reference at the beginning of the process. Being aware that weighting scales of different brands can give different numbers, the person buys three scales, colored in blue, in orange and in green. The scales are of the mechanical type, with levers and springs.

1. ppm: part per million.
2. This is not the place to explain in details the difference between mass and weight. It is assumed here that a person with a 190-lb mass has a weight of 190 lbf. However, the gravitational constant that explains the variations of weight with the same mass is shortly discussed in paragraph 2.4.

The blue scale indicates 188 lbf, the orange 194 lbf and the green 196 lbf. Confused and depressed by the numbers, the person tries again. Now, the blue scale announces 190, the orange 191 and the green 195. Trying to make sense of all these numbers, the person observes that he wears clothes, and among them, a belt with a heavy metal buckle. He decides to strip off his clothes. In addition, for completeness, he removes his cheap and light watch. Now the blue scale indicates 183 lbf, the orange 187 lbf and the green 189 lbf. He draws Table 2.1.

Table 2.1 Bathroom scale experiment. Unit is lbf.

All weights in pounds	**Blue**	**Orange**	**Green**
First measurement	188	194	196
Second measurement	190	191	195
Third measurement	183	187	189

Considering the objective, weight loss, what is the strategy that should be followed? Some observations must be made:

- The measurements range from 183 lbf to 196 lbf, a difference of 13 lbf.
- Between the first and the second measurements, the smallest difference, one lbf, is observed with the green scale.
- It is intuitive that the third set of measurements is more reliable as it offers the most control. If several measurements are taken successively, it would be preferred to remove the variation in clothing (and belt selection) and involve only the body in the weighting experiment.

Wiser with the previous observations, and unknowingly becoming a better metrologist, the man decides to monitor his weight with a process including the following restrictions:

- He will use the green scale to perform the successive measurements, and at a lesser frequency, will do checks with the blue scale in order to verify that the green scale functions consistently.
- Every time he wants to be informed on his weight, he will perform three measurements. If they do not vastly differ, he will take the average of the three. Otherwise he will investigate the reasons of the larger spread.
- He will ensure to be without clothes, or if constrained to do otherwise, will wear exactly the same clothes. He understands that wearing clothes will result in a less satisfactory measurement.
- He understands that a variation of one lbf in weight is not significant. A reduction of one lbf cannot be considered as a success, or a failure of the diet.
- He understands that he can use other weights for reference. He buys a set (Fig. 2.2) and uses them to calibrate the bathroom scale. He gets a table of correspondence between the values indicated by the bathroom scale and the values shown on the commercial weights (Table 2.2). For intermediate values, he uses interpolation.
- He understands now that the scale measures weights through springs, metal bars and other gizmos. He will search for layman information on internet about scale design and building. He eventually will contact the scale manufacturer to obtain the specifications of the scales he has purchased.

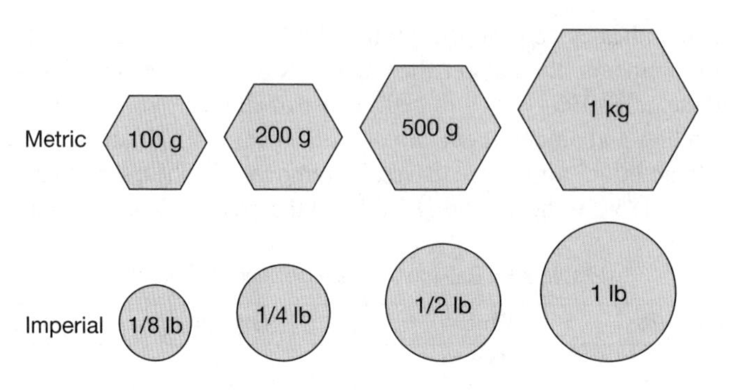

Figure 2.2

Set of commercial weights.

Table 2.2 Comparison of calibrated values and raw values

	Weights (lbf)	Bathroom scale (lbf)
1	100	98
2	150	149
3	200	203

2.3.1 A minimalist mechanics course on how a bathroom scale works

Fig. 2.3 is a simplified representation of a bathroom scale. Fig. 2.4 shows how a spring stretches under a tension pull [1].

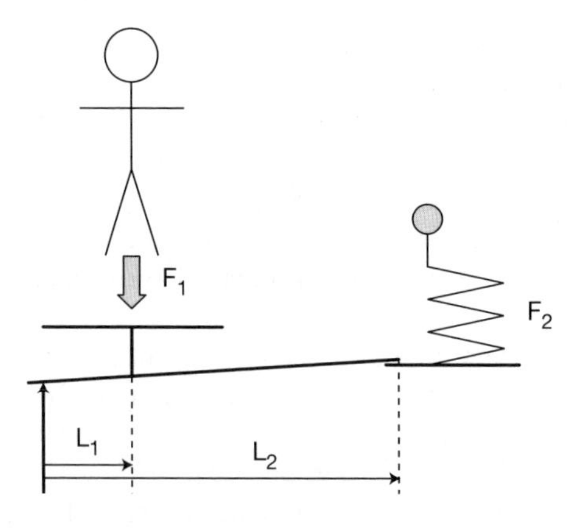

Figure 2.3

How does a bathroom scale work?

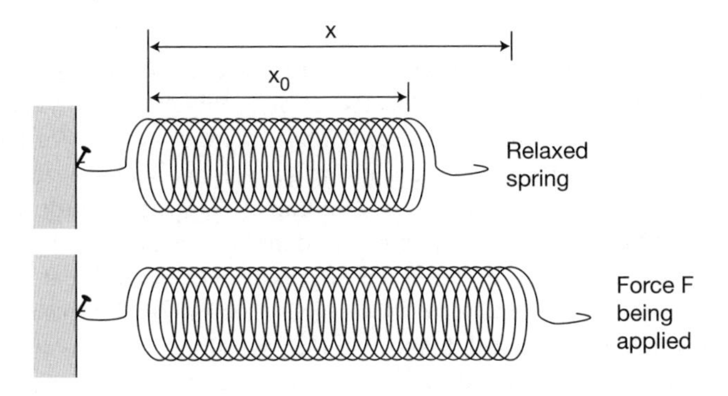

Figure 2.4

Change of spring length with applied force.

Most bathroom scales use a lever rotating around a fixed point. The weight of the user creates a moment M_1 as the weight F_1 acts at the distance L_1 of the extremity of the lever. The lever causes a spring to extend by a length $x-x_0$. The following equations describe what is happening:

- Weight: $F_1 = mg$
- Equality of moments: $M_1 = F_1 * L_1 = F_2 * L_2$
- Stretch of the spring: $F_2 = k (x - x_0)$

m is the mass of the person to be weighted,
k is the stretch constant of the spring,
x_0 is the stretch of the spring with no additional weight on the scale,
L_1 is the distance between the fixed point and the plateau,
L_2, is the distance between this point and the attachment to the spring,
g is the gravitational constant.

The stretch of the spring, $x - x_0$, is connected to a dial through some additional mechanical link. The complicated gears between the spring and the dials are not investigated here. x is assumed to be the parameter indicating the weight.

The final equation yielding the weight is:

$$m = f(x) = F_1/g = F_2\ L_2/(L_1\ g) = k\ (x - x_0)\ L_2/(L_1\ g)$$

2.3.2 Sensitivity of the measurements to parameters

There are five parameters involved in the measurement of this simplified bathroom scale, L_1, L_2, k, x_0 and g. A sturdy design of the bathroom would make L_1 and L_2 reasonably constant.

Parameter x_0

With age and use, x_0 is likely to increase. The change of the value of x_0 will be compensated by recalibration of the bathroom scale. In any case, this evolution is slow and may not affect the user on the short duration of the diet.

Parameter g

The strength of the gravitational field at the earth's surface, denoted g, is approximately equal to 9.81 m/s^2 or 32.2 $ft/s.^2$ Since the local force of gravity can vary by up to 0.5% at different locations, spring scales may measure slightly different weights for the same object (the same mass) at different locations. To standardize weights, scales need to be calibrated to read the weight an object would have at a nominal standard gravity of 9.80665 m/s^2 (approximately 32.174 ft/s^2). This calibration is done at the factory. When the scale is moved to another location on earth, the force of gravity will be different, causing a slight error. So, to be highly accurate, and legal for commerce, spring scales must be re-calibrated at the location at which they are used. In the extreme case of measuring weight on the moon, the gravity constant is about one-sixth of what it is on earth or 1.625 m/s^2. The measurement, calibrated on earth would be incorrect by 83%.

Parameter k

For most objects, the "Young's modulus", that would control the change of k, changes with temperature, as they are more easily stretched as temperature increases. Therefore the spring constant will get smaller as the temperature goes up.

The data flow relating the stretch of the spring, the geometrical characteristics of the device, the local gravity constant and the final weight is shown in Fig. 2.5.

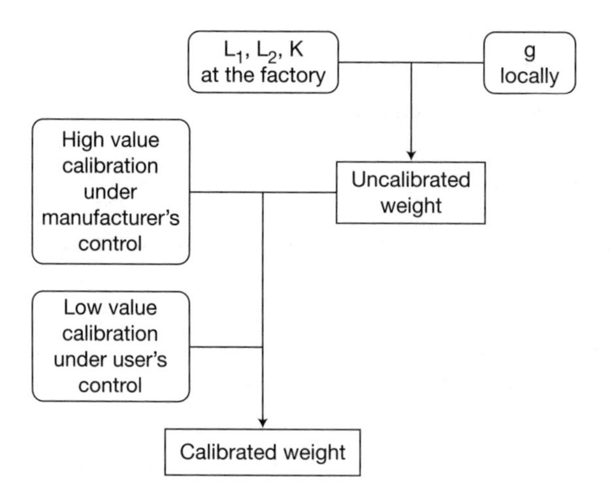

Figure 2.5

Flow chart of the bathroom scale measurement.

2.3.3 The pleasing bathroom scale

A company manufacturing bathroom scales could include a small computer and some software to please the customer and show a steady decrease in weight with time regardless of the real values of the weight. The bathroom scale would not deliver a measurement close to reality, but a measurement designed to please the user. Similarly, some companies in the oil-field are more concerned about delivering pleasant news than accurate ones.

2.4 COMBINING MEASUREMENTS AND HOW A CALCULATOR CAN BE DECEIVING

Our user is more interested with his Body Mass Index, BMI, than by his weight. BMI is defined as:

BMI = weight/(height*height),

where the weight is expressed [3] in kg and the height in m.

He takes 189 lbf from the green scale, makes the conversion to the metric system, 85.9 kg, and measures his height with a coarsely sampled tape, which gives an indication almost right in the middle of two "cm" indications: 1.685. Taking his hand calculator (Fig. 2.6), he computes: 30.254734. In fact considering that the weight is between 175 (79.5 kg) and 190 lbf (86.2 kg) and that the height is between 168 and 169 cm, the BMI is an unknowable value between two knowable limits, 27.8 and 30.6. The 10 figures obtained from the calculator give an unreasonable feeling of high precision. [4]

Figure 2.6

Picture of the calculator showing the BMI computation. The numerous figures give an erroneous sense of high precision.

2.5 COMPARISON WITH OIL INDUSTRY MEASUREMENTS

At the beginning of the 21st century, the most usual process used by the oil company is to collect digital data from a data acquisition company. The data is then loaded in an interpretation software in order to derive rock properties. A log digital file is shown in Table 2.3.

3. There is a multiplier of 703 if the weight is in pounds (lbf) and the height is in inches.

4. Interestingly enough, a value of 27.8 qualifies the user as "overweight" while a value of 30.6 brings him in the "obese category." See Chapter 11 on decision thresholds.

Table 2.3 log data shown in LAS format

Depth	Δt	Bit size	Caliper	Density	Drho	GR
m	µs/ft	in	in	g/cm^3	g/cm^3	GAPI
581.7180	118.352	9.5000	9.6518	2.0929	0.0416	46.4436
581.8704	118.192	9.5000	9.6305	2.0929	0.0416	47.4542
582.0228	117.941	9.5000	9.6305	2.0885	0.0339	46.5120
582.1752	117.300	9.5000	9.5984	2.0947	0.0232	46.5703
582.3276	116.310	9.5000	9.6198	2.1052	0.0162	45.3280
582.4800	115.642	9.5000	9.6198	2.1274	0.0281	44.6865
582.6324	115.107	9.5000	9.6269	2.1213	0.0298	44.7658
582.7848	114.232	9.5000	9.5771	2.1269	0.0357	44.4736
582.9372	113.260	9.5000	9.5664	2.1214	0.0279	45.8879
583.0896	112.215	9.5000	9.6518	2.1325	0.0320	44.5343

How is this data different from the one collected in the bathroom scale experiment?

- The numbers representing the measurements have too many significant figures, that is, figures after the point (four except for Δt in the example above). Consider the depth shown on the third line. It is expressed with a resolution of 0.1 mm or about four thousandths of an inch. Technology is totally unable to describe depth so well. Similarly, consider the density 2.0929 g/cm^3, shown on line 3, column 5. It is not possible to design equipment that would yield such detailed information. In other words, the logging information is delivered in a misleading way that confers an unrealistic sense of sharpness.
- While measurements performed by professional metrologists include a range of values to the user, only one value is listed here for each depth. The information is definitely not crosschecked with a different set of equipment. A second measurement with the same tool is sometimes performed in the context of a "repeat section," but this repeat section is seldom used in the interpretation process.
- The logging measurement is not optimized to suppress the undesired signals (the clothes in the bathroom scale experiments). For instance, borehole effects affect most measurements and are always present.

2.5.1 Unfounded feeling of accuracy and precision on derived properties

A simplistic approach in interpretation is to directly compute formation characteristics from measurements:

Measurements

Measured porosity $\phi = 21$ pu,

Measured formation resistivity $R_t = 28$ ohm-m,

Formation water resistivity: $R_w = 0.12$ ohm-m,

The saturation is assumed to be: $S_w = (1/\phi)\sqrt{R_w/R_t}$ and is computed as $S_w = 31.1739843$. In reality, the measurements are not perfect and are affected by uncertainties:

Uncertainties [5]

$\sigma_\phi = 2$ pu

$\sigma_{Ct} = 1$ mS/m

$\sigma_{Rw} = 0$

The propagated uncertainty [2] is: $\sigma_{Sw}/S_w = 9.6\%$ or $\sigma_{Sw} = 3$ su [6]

As a result S_w is found between 28.2 and 34.2. So, not even the second number (1) of the original computations is certain.

2.5.2 Difference between rock properties and logging measurements

Considering the caution to be taken to use bathroom scale weight, a measurement performed at surface and in an easily accessible environment, it is necessary to analyze what may affect the information collected miles underground in conditions that cannot be observed.

In the next three chapters, it will be shown that:

- All logging measurements are using indirect processes to get close to rock information.
- No logging measurement can be focused to only look at the reservoir rocks or zones of interest.
- Similarly to the spread observed in the scale experiment, multiple values are obtained when logging is repeated. In addition, all these values differ from the true value of the formation.

2.6 SUMMARY

- Reality cannot be easily grasped. Observations are biased by the limits of our senses.
- Measurements enable a quantitative description of reality, but it is approximate.
- A simple measurement of the length of a rod does not give a single answer.
- Deriving weight from a ruler or from a bathroom scale is not as simple as it looks.
- When different measurements, such as length and weight, are combined to take a decision, care must be exercised.
- Calibrating the measurement with a known reference increases the value of the measurement.
- Logging measurements are also quite complicated.

---

5. Chapters 12 and 13 inform the reader on how to derive these numbers.
6. One su is one saturation unit.

- A digital record of well measurements gives a unique and extremely sharp value, which detracts from reality.
- Note: In order to build the graphical examples on length and weight, "true" values of 29.4237 (11.5838 in.) and 179 lbf (81.2 kg) were used. The reader is invited to compute the differences between these "true" values and those derived from measurements.

REFERENCES

[1] http://home.howstuffworks.com/inside-scale.htm

[2] Theys, P., *Log data acquisition and quality control*, Éditions Technip, 1999.

3

All well measurements are indirect

The limitations of physics are such that the underground formations are not measured directly. The most common underground measurements are reviewed. A strong emphasis is given to resistivity as recent technological developments have improved this difficult measurement. A summary of the logging process is given first.

3.1 A BRIEF DESCRIPTION OF THE LOGGING PROCESS

There are two ways to collect information on underground rocks and fluids.

1) Take a piece of these rocks, bring it to surface and perform measurements on the sample. Coring [1] is the name of this method. The limitations of coring are shortly reviewed in Chapter 19.
2) Attach nuclear, electrical, acoustical, etc. sources to a cable or to a drill string, add also sensors, excite the underground formations using these sources, collect the resulting signals with the sensors, and transmit these signals through the cable – which has electrical conductors – or through the mud column. [2] In some cases, there is no need to stimulate the formations as they generate signals spontaneously. The later measurements are qualified as passive.

3.2 SPONTANEOUS POTENTIAL

Spontaneous potential, or SP, is the second historical downhole measurement. It has applications in the distinction of porous and impervious rock beds and in the quantification of the differences between formation and mud-derived fluids. The electrical circuit of SP is simple.

1. Carottage, in French.
2. The process is analogous to the use of a flash device that illuminates a person. A picture can then be taken to describe that person.

A very low current is measured through a very large resistor. In addition there are a few resistors and a battery. The measurement, though simple, can be affected by a number of parasitic signals, such as cable magnetization and stray voltages.

The results of the interpretation of SP, shale content V_{sh} and formation water resistivity R_w, involve a number of parameters that makes the method indirect. A proper choice of these parameters is a prerequisite.

3.3 RESISTIVITY

Resistivity is the first downhole measurement. Logging started with this formation characteristic because it displays a huge contrast (from 0.1 ohm-m to 100,000 ohm-m) between water and oil, between saline and fresh waters, between porous and hard formations. The measurement of a parameter with such a dynamic range does need to be extremely accurate to be useful.

This historical debut is a paradox, as the quantification of formation resistivity is an exceedingly challenging proposition. Eighty years after the first resistivity run, new tools attempting to measure formation resistivity are still being designed and developed. In fact R_t, the resistivity of the formation far away from the borehole is what the petrophysicist is looking for, while resistivities are actually measured at some points in the borehole. Henri Doll, [3] in the 1970s, confessed [2] that, for more than 30 years, the resistivity measurements had little or no connection with R_t.

3.3.1 Early tools

On the first log, run in 1927, the curve was named resistivity, not formation resistivity. This can be easily explained as the key component of the instrumentation was a surface potentiometer (Fig. 3.1). It functions with dials of resistivity values. A single parameter (the K factor of the downhole sonde) is involved before the potentiometer resistivities are plotted on a graph.

After a few years of experimentation, the tools have gained electrodes or coils and can function in different modes. In the 1930s, the shape and amplitude of the curve depended on the spacing between the electrodes. Many charts [4], trying to solve for invasion, borehole and changing bed property effects were needed to arrive close to formation resistivity (Fig. 3.2). At this early stage, inverse models were already used. The chart does not indicate the values of R_t from the measurements, but link the measurements to R_t.

3. See Chapter 17.

Figure 3.1

Schlumberger potentiometer, 1925.
By courtesy of Dominique Chapellier.

BED THICKNESS (e)	QUALIFICATIONS	DEVICE	RESPONSE
A. IN LOW RESISTIVITY, WHEN $R_{16"}/R_m < 10$ (INVASION UP TO 2d)			
$e > 20'$ (>4 AM')		Long Normal	$R_{64"} = R_t$
$e \simeq 15'$ (3AM')	$R_m \simeq R_s$ $R_{64"}/R_s \geq 2.5$	Long Normal	$R_{64"} = 2/3\ R_t$
$e \simeq 15'$ (3AM')	$R_m \simeq R_s$ $R_{64"}/R_s \leq 1.5$	Long Normal	$R_{64"} = R_t$
$e \simeq 10'$ (2AM')	$R_m \simeq R_s$ $R_{64"}/R_s \geq 2.5$	Long Normal	$R_{64"} = 1/2\ R_t$
$e \simeq 10'$ (2AM')	$R_m \simeq R_s$ $R_{64"}/R_s = 1.5$	Long Normal	$R_{64"} = 2/3\ R_t$
$5' < e < 10'$	When oil bearing and SP is -50 – 80 MV	Short Normal	$R_{16"} \simeq R_t$
$5' < e < 10'$	Surrounding beds homogenous	Lateral in resistive bed	$R_t \gtrsim R_{Max} \times R_s/R_{Min}$
Thin beds (in general)	Surrounding beds homogenous	Lateral in conductive bed	$R_{19"} \simeq R_t$

Figure 3.2

Table linking the measurements to formation resistivity. The index of the curve is the spacing (e.g., 64 in.). The measurements are expressed as functions of R_t.
By courtesy of Schlumberger.

3.3.2 Recent tools

No modern so-called resistivity device actually measures resistivity,[4] but voltages and intensities. Many tools from many vendors are available and the principles are not explained here in details. Cutting the chain of processing short, it is accepted that there are two broad categories of devices:

- The electrode tools (mostly sensitive to resistivity), also known as laterologs.
- The induction tools (mostly sensitive to conductivity).

A practical example (Fig. 3.3) sheds some light on resistivity logging [5]. A well is drilled through separated beds with R_t resistivities varying from 0.3 ohm-m (9,920 up to 9,880 ft), through 1 ohm-m (9,880 to 9,850 ft) to 90 ohm-m (9,850 to 9,800 ft).

The formation water resistivity, R_w, the porosity ϕ and the formation factor F are constant. A simple Archie saturation equation is used:

$$R_t = R_w/(\phi^2 * S_w^2).$$

The resulting water saturation profile has successive sharp steps, 100 su in the lower zone, 55 su in the intermediate (transition) zone and 5.8 su in the upper zone (hydrocarbons). The invaded zone resistivity, R_{xo}, goes from 1.8 ohm-m to 2.2 ohm-m then 4 ohm-m. The diameter of invasion is selected in the model as 50 in., a reasonable assumption for a wireline job for which the borehole has been open for a few days.

The true (modeled) and measured values obtained by laterolog and induction methods are gathered in Table 3.1.

Table 3.1 True and measured values for the curves shown in Fig. 3.3.

Zone	Depth (ft)	Depth (ft)	R_t ohm-m	R_{xo} ohm-m	S_w (su)	R_{LL} ohm-m	R_{LL}/R_t (%)	S_{wLL} (su)	S_{wLL}/S_w (%)
1	9,920	9,880	0.3	1.8	100.0	1.0	333.3		
2	9,880	9,850	1.0	2.2	55.0				
3	9,850	9,800	90.0	4.0	5.8	42.0	46.7	15	265.5

Zone	Depth (ft)	Depth (ft)	R_t ohm-m	R_{xo} ohm-m	S_w (su)	R_{ID} ohm-m	R_{ID}/R_t (%)	S_{wID} (su)	S_{wID}/S_w (%)
1	9,920	9,880	0.3	1.8	100.0	0.4	126.7		
2	9,880	9,850	1.0	2.2	55.0				
3	9,850	9,800	90.0	4.0	5.8	30.0	33.3	11	194.8

4. Even a surface ohm-meter does not measure resistivity directly.

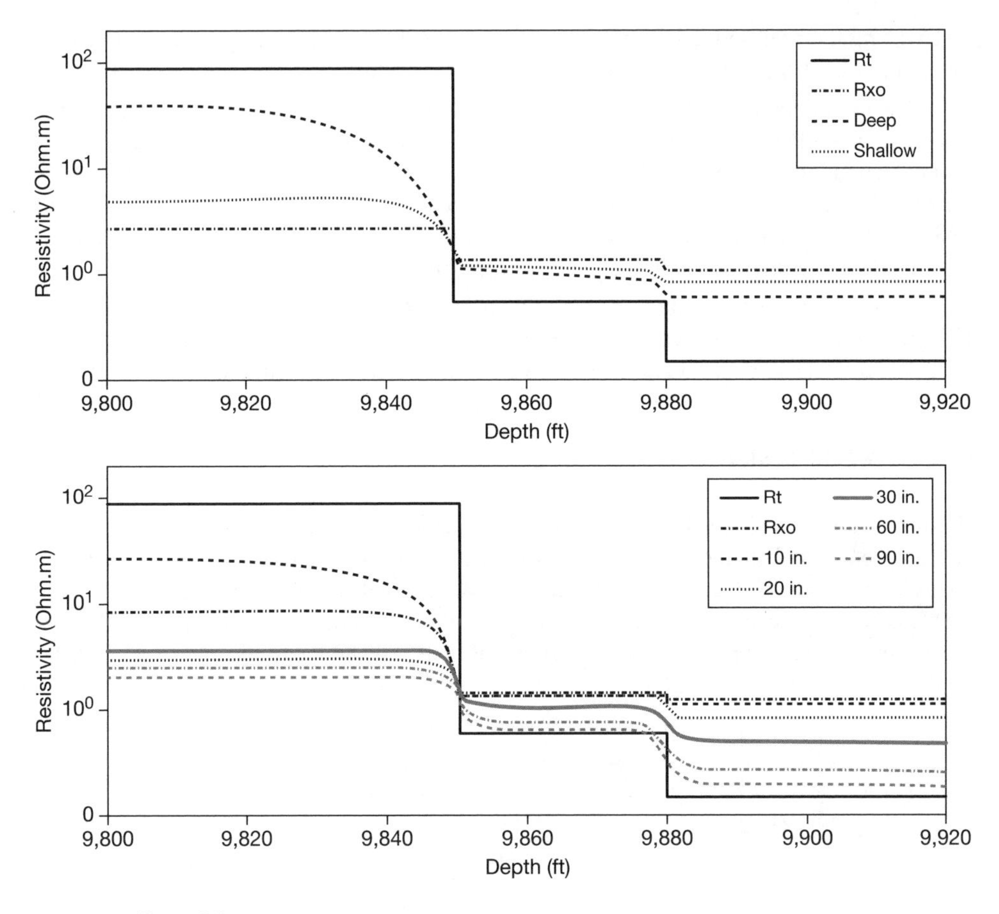

Figure 3.3

Comparison of true values and measured values. Large differences between the model (sharp looking lines) and the measured curves (softer transitions) are obvious.

By courtesy of SPWLA.

Laterolog readings

The deep (long spacing) laterolog R_{LL} reads 1 ohm-m in the water zone, in other words 3.3 times more than the true value, R_t. R_{LL} reads 42 ohm-m in the hydrocarbon zone, or 2.1 times less than the true value. As a consequence, the minimum saturation in the hydrocarbon zone is 15.4 su or 2.7 times larger than the true value.

Induction readings

The deep induction (90 in.-spacing), R_{IL}, reads 0.38 ohm-m in the water zone, 27% more than the true value. The same curve reads 30 ohm-m in the hydrocarbon zone, three times less than the real value. The resulting minimum water saturation in the hydrocarbon zone is 11.3 su or twice the true value.

By using the deep induction in the water zone and the deep laterolog in the hydrocarbon zone, saturation becomes 9.5 su or 65% above the real value.

It is worth noticing the transitory sections on all curves as the true values go sharply from one value to another. The longest transient intervals are observed on the deep curves (12 ft on the deep induction, 15 ft on the deep laterolog). These buildups could be interpreted as a transition zone with variable water saturation while the real "transition" bed is between 9,880 and 9,850. Misinterpretation on the transition zone may result in the wrong evaluation of the thickness of the reservoir and hence of the volume of hydrocarbon in place.

Finally, R_{xo} is better evaluated using the 10 in. induction and bed boundaries are well defined by this curve. Through this more accurate definition of the bed boundaries a correct formation model can be built (forward modeling) and accurate saturations can be computed.

3.3.3 Tri-axial resistivities

Multi-axial resistivity tools have been recently developed. The measured values are still not strictly identical to formation values. This is shown in Fig. 3.4 where conductivities are plotted versus r_1, the radius of invasion. The true formation conductivity is the horizontal line at 1,000 mS/m. All the measurements are sloping lines. It is only when invasion is minimal, which happens in non-productive shaly formations, that real and measured values coincide [1].

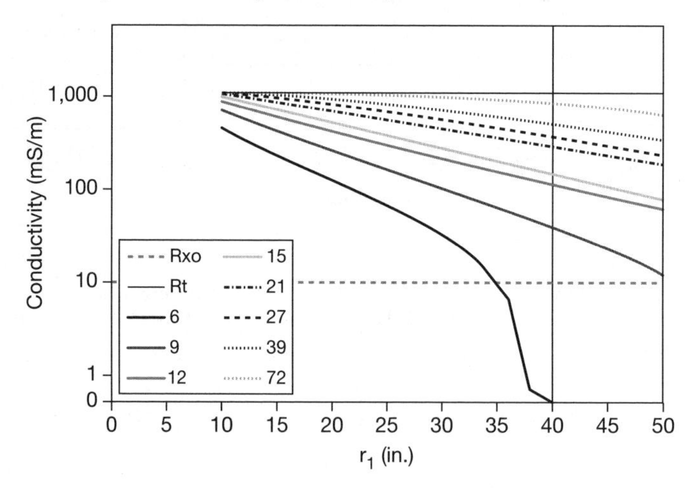

Figure 3.4

Response of the different spacings. r_1 is the radius of invasion.
By courtesy of SPWLA.

3.3.4 Exploiting the difference of response between resistivity tools

Differences of response between the resistivity tools can be used to detect anomalies in the formation. Fractures are seen by laterolog-type tools, but do not affect induction-type tools.

A combination of the two types of measurements enables the recognition of fractures (Fig. 3.5).

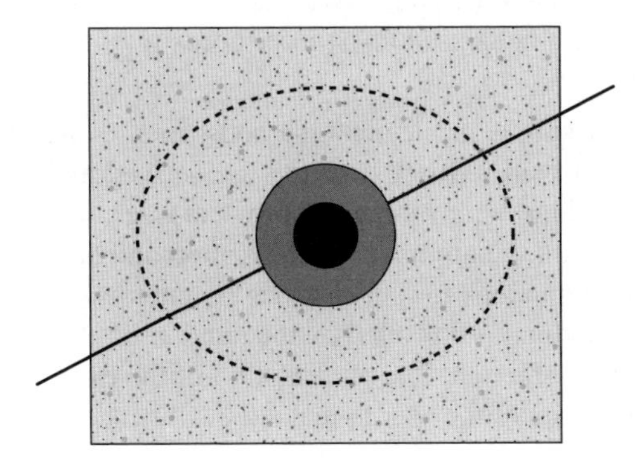

Figure 3.5

Difference of current patterns of a laterolog and of an induction type tools enable fracture detection. The current pattern of the induction is a circle. The current pattern of the laterolog is a straight line. The fracture invaded by a conductive mud dramatically changes the laterolog reading as compared to a tight (highly resistive) formation.

In a similar way, logging devices designed with different physics principles can help explore the special features of the rocks. The neutron tool, reviewed later, can be built to detect epithermal and thermal neutrons. The difference of response can help identify the presence of neutron absorbing minerals.

3.4 GAMMA RAY

Natural radioactivity enables the typing of rocks. This is why early logging included a measurement of the number of gamma rays emitted spontaneously by the geological formations.

The gamma ray measurement appears as quite simple. A gamma ray detector is lowered in the well and gamma rays received in a given time interval are counted. The first gamma ray tools used to be scaled in cps, counts per second. Difficulties arise as the size, efficiency and attenuation in the tool housing need to be somewhat taken in account. This is done by converting the counting rates in API units, a process that is tool and formation dependent. In spite of this alignment to a common standard, data users are often complaining about differences observed when two different tools are used. Fortunately, the gamma ray curve is often used in qualitative interpretation to determine the percentage content of shale. The spectral gamma ray tool (giving useful information on Thorium, Potassium and Uranium concentrations) involves complex window and filtering processing. In that case, a complicated calibration is required as the concentrations are then used in a quantitative approach.

3.5 DENSITY

Density is an important formation characteristic as it varies with rock type, fluid type and fluid content. This measurement is used to estimate porosity, the proportion of the total volume occupied by fluids.

It is not possible to use simple mechanical methods, such as the ones described in Chapter 2, to derive downhole densities. Rock densities are derived indirectly through the analysis of gamma ray interactions with matter. Gamma rays are emitted from a radioactive source (Cs 137), interact with the atoms of the formation and are detected some distance a few inches away. Two interactions may take place: Compton scattering and photoelectric absorption. Compton scattering is controlled by the number of electrons in the formation, or the electron density of the formation. The presentation of the early first density tools was in cps and not in g/cm^3. It was up to the user to make the conversion (Fig. 3.6).

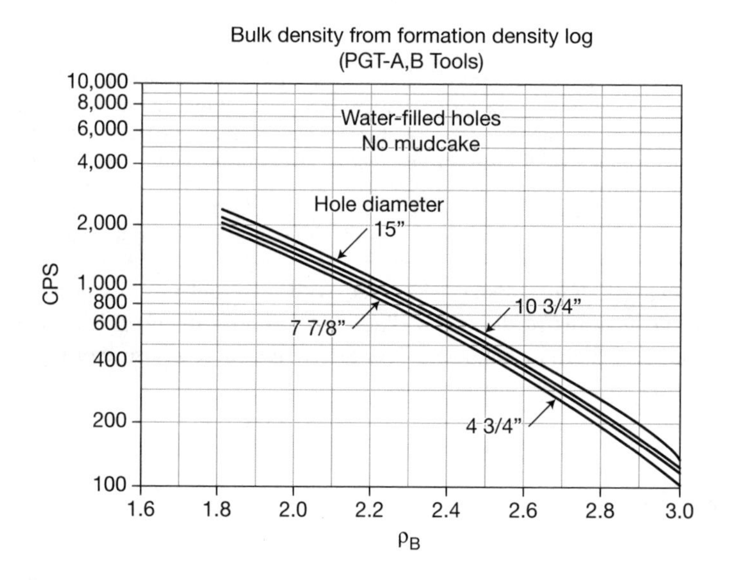

Figure 3.6

Original presentation included a density-related curve scaled in cps.
By courtesy of Schlumberger.

3.5.1 Electron density

The inconvenient cps scale was soon replaced by the derivation of electron density. It is performed by the analysis of the Compton-scattered gamma rays corrected for photo-electric absorption at low gamma ray energies. In the design of the early tools, low energy gamma rays were blocked by cadmium shields mounted on the detectors. In later modern designs, the separation of low energy gamma rays from those not affected by photo-electric absorption is performed by spectral analysis.

3.5.2 Density

The electron density is still not the same as bulk density. In water-bearing limestones, it is possible to go from one to another by a simple transform: [5]

$$\rho = a\,\rho_e - b, \text{ where } a = 1.0704 \text{ and } b = 0.188$$

For other minerals, some differences are observed, but they are completely predictable, once the mineral is identified (Fig. 3.7).

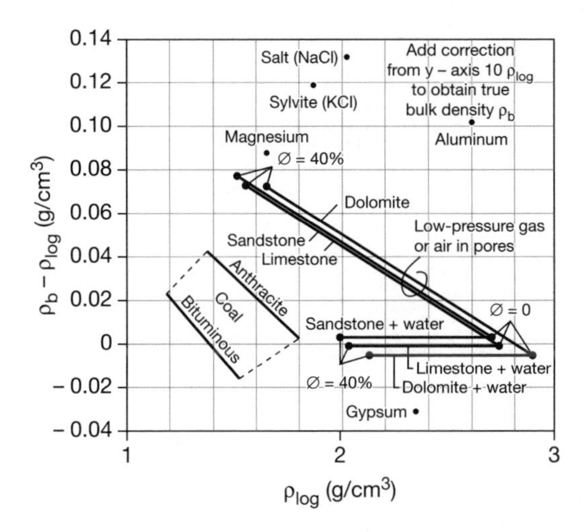

Figure 3.7

Transform between limestone-calibrated densities and real densities.

Enter the chart with the log density reading and move upward to intersect the mineral line of interest. Example: The log reads: 2.05 g/cm^3 and the cuttings indicated that the well crosses a thick salt bed. So, from the 2.05 g/cm^3 point on the x-axis, it is possible to move up to the salt point and read 0.135 g/cm^3. The real density is 2.185 g/cm^3, which is the correct value for the density of salt.

By courtesy of Schlumberger.

3.6 COMPRESSIONAL SONIC

The acoustic properties of underground formations are measured to enable correlation with surface seismic and to provide an estimation of fluid content.

It is possible to measure the time taken by a compressional acoustic wave between a transmitter and a receiver in a tool. This wave travels not only through the nearby formation, but also through the mud column separating the tool from it. A burst of acoustic energy is

5. This transform is explained in details in reference [3].

emitted from the transmitter, and the first arrival of the compressional wave is detected at the receiver (Fig. 3.8).

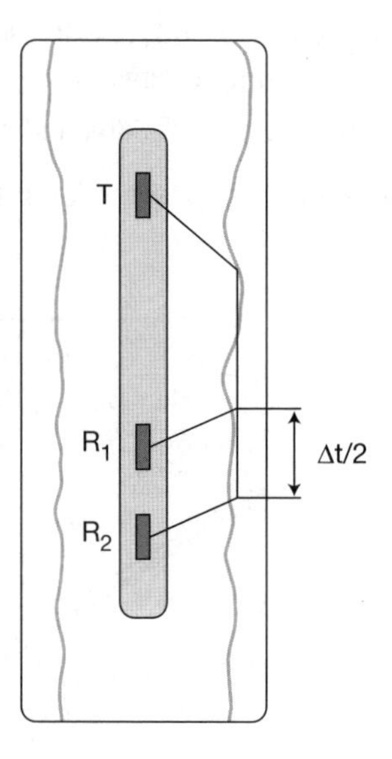

Figure 3.8

Simplified principle of the measurement of a compressional waveform. T is the acoustic transmitter. R_1 and R_2, the acoustic receivers. The transit time along the formation is divided by 2 because the spacing between receivers is 2 ft. The measurement is expressed in μs/ft.

3.7 OTHER SONIC INFORMATION

If the timing of the first arrival of an acoustic wave is simple, the other acoustic measurements cannot be obtained readily. Complex processing, such a slowness-time coherence analysis may be required. For dipole sources tools, the shear wave is not even measured at all. The flexural mode is used instead to indirectly measure the shear mode. Measurement of the flexural mode is the only technique available in slow formations where shear velocity is less than the borehole-fluid velocity. The flexural mode is closely related to the shear mode, but unlike the shear mode, it has strong frequency dependence. Thus, additional signal processing is needed.

3.8 NEUTRON LOGGING

The analysis of neutron interactions with geological formation has shown correlation with fluid content and with water saturation. The first "neutron" type devices were displaying counting rates curves (Fig. 3.9) and much skill was needed to arrive to some kind of porosity [4].

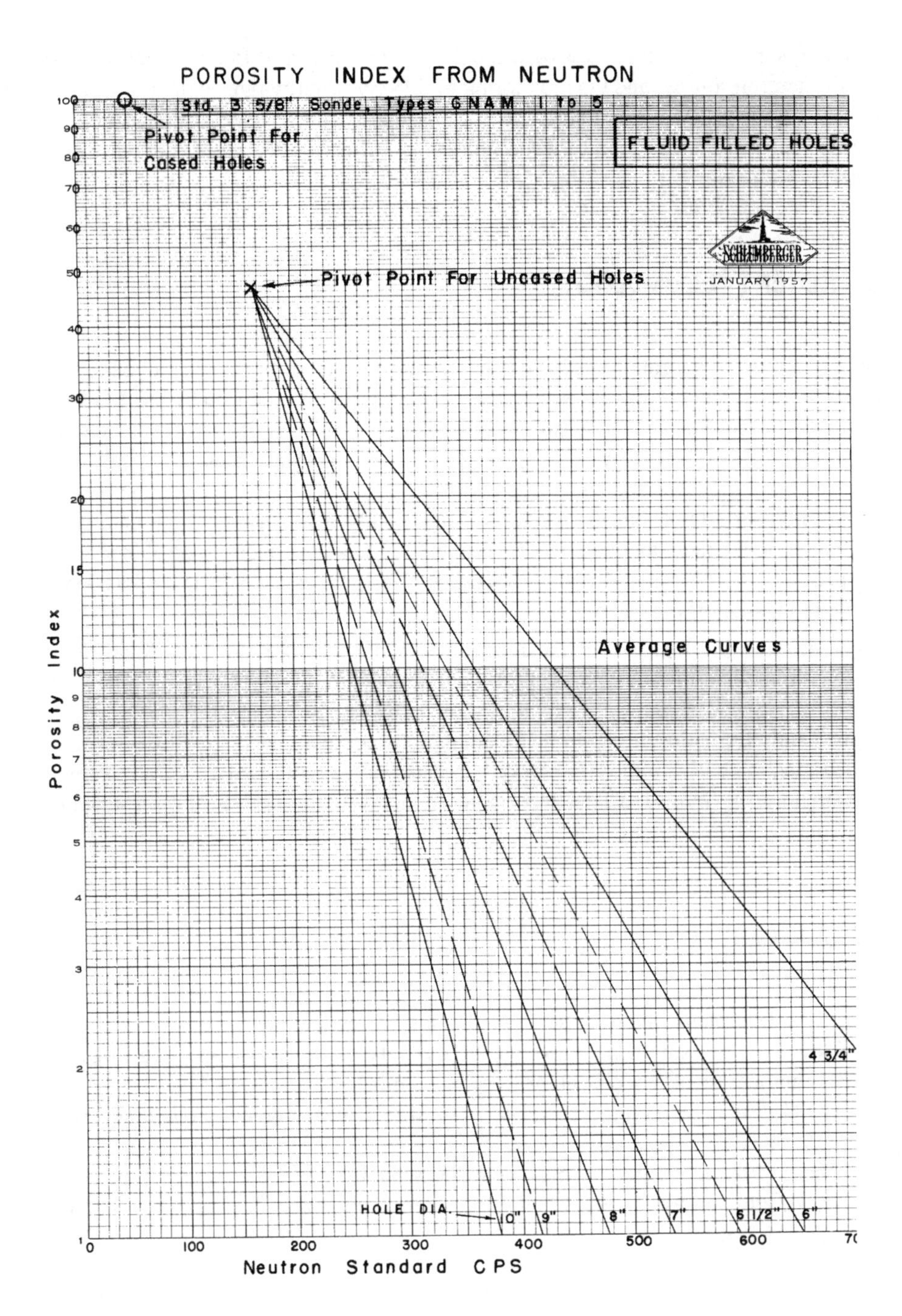

Figure 3.9

Early neutron chart.

By courtesy of Schlumberger.

Subsequent neutron designs involved multiple detectors at different spacings (distances from the neutron source) and the monitoring of different energy ranges (thermal and epithermal). The transform from counting rates to porosity is established through considerable experimental work with laboratory formations and after application of numerous environmental corrections.

3.9 RECENT TECHNOLOGY

A small number of logging tools have been briefly described. In the recent years, the introduction of newer technology has been massive. The raw data produced by the new tools is in fact similar to those produced by earlier tools. The difference is that multi-spacing or multi-detector devices are now common. Spectral analysis is also often used. Signal processing has become complicated and cannot be understood by the average user of the data. This is not the data user's fault, but the vendor's. The logging company owes the data user a full explanation on the new devices and processing. The data user needs to be able to access the information, even if it is simplified for his usage. Black boxes and arcane models do not help in the optimization of the use of data.

The demarcation line between acquisition and interpretation becomes fuzzy. When the term "neutron counting rates" is used for a curve, there is no doubt that this is not a formation porosity that is displayed. Conversely, the expressions "neutron porosity," "formation density," and "formation resistivity" are ambiguous and may encourage the user to believe that he is dealing with a directly measured porosity, density or resistivity void of any degree of uncertainty. As indicated later in Chapter 9, many terms used in log presentations are compounding the misunderstanding.

3.10 SUMMARY

- All well measurements are performed in an indirect manner.
- Downhole measurements make use of a variety of physical laws, requiring a number of computational models and processing parameters.
- Calibrations are required to relate the indirect measurements to the actual desired properties.
- These different parameters must be made available to the data user.
- Packing the physics used in tool design into black boxes does not facilitate the work of the data users.

REFERENCES

[1] Barber, T., Wang, H., Leveridge, R., Hazen, G., Schlein, B., "Principles of log quality control for complex induction logging instruments," paper MMMM, Trans. SPWLA 49th annual logging symposium, 2008.

[2] Bowker, G., *Science on the run, information management and industrial geophysics at Schlumberger, 1920-1940*, The MIT press, 1994.

[3] Ellis, D. V., Singer J. M., *Well logging for earth scientists*, Springer, 2007.

[4] Schlumberger, *Historical charts*, SMP-7030.

[5] Tabanou, J., Theys P., "Le log," *The log analyst*, 1992.

4

Logging measurements do not focus on zones of interest

When a human being uses a bathroom scale, a few external elements may interfere with the measurement: the clothes and the surface (carpet or tile) on which the scale is laying on. In a well logging measurement, it is obviously impossible to isolate an underground piece of rock and assess its properties. [1] In addition, the geometrical characteristics of the formations of interest may considerably vary: They could be massive and thick or tiny and thin. Conversely, the geometry of a given logging tool is frozen when it is commercially launched. Distances between transmitters and receivers, sources and detectors do not change. The mismatch between varying rock properties and fixed logging tools designs is a considerable challenge in the quest of values representing reality.

4.1 VOLUMES INVESTIGATED BY LOGGING TOOLS

4.1.1 Open-hole conditions

Logging tools are designed to survey zones of interest, potential reservoirs or distinctive markers. [2] It is impossible to prevent them from having a peek at what is around. In Fig. 4.1, representative of an open-hole configuration, the volume of investigation of the ideal logging sensor is shown on the top. The sensor ignores the borehole, the invaded zone, and the neighboring zones. It focuses on the virgin zone unaffected by drilling and invasion. Such a sensor cannot be designed. A real logging sensor would generally look at:

- the borehole,

1. Coring collects downhole samples. Measurements performed on cores have their own limitations. Refer to Chapter 19 for more details.
2. The same applies when casing, tubing and other downhole objects are under surveillance.

- the invaded zone,
- the beds above and below, also called shoulder beds.

Fig. 4.2 gives a detailed lateral view of the borehole and of the surrounding zones, with the numerous parameters linked to them.

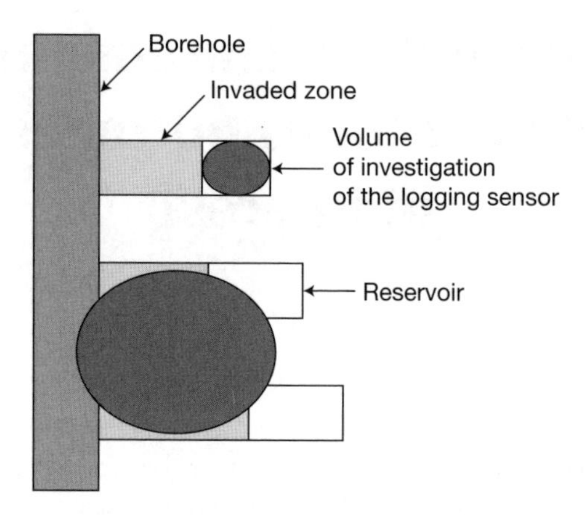

Figure 4.1

View of a vertical well. The black area represents the volume of investigation of a logging tool. The top bed is surveyed by an "ideal" sensor concentrating on the reservoir. The lower beds are surveyed by a more realistic sensor, affected by the borehole, the invaded zone and the successive beds.

4.1.2 Cased-hole conditions

Cased-hole environments can be even more challenging. Because of size restriction in the tubings, logging tools have small external diameters. In the situation shown in Fig. 4.3, the logging tool, shown in black, is located in one of the two production tubings. Production fluids are different in those two. A third type of fluid is present in the casing. Tubing and casing, generally made of steel also contribute to the total signal, in competition with the formation signal. Cement is not represented, for clarity, but its impact needs to be also accounted for. The cement volume and shape do depend on the original open hole that was drilled. The volume of investigation of the logging tool is represented by a dashed circle. It is intuitive that only a portion of the total signal relates to the formation. A careful evaluation of these non-formation contributions is required before a successful interpretation can be completed.

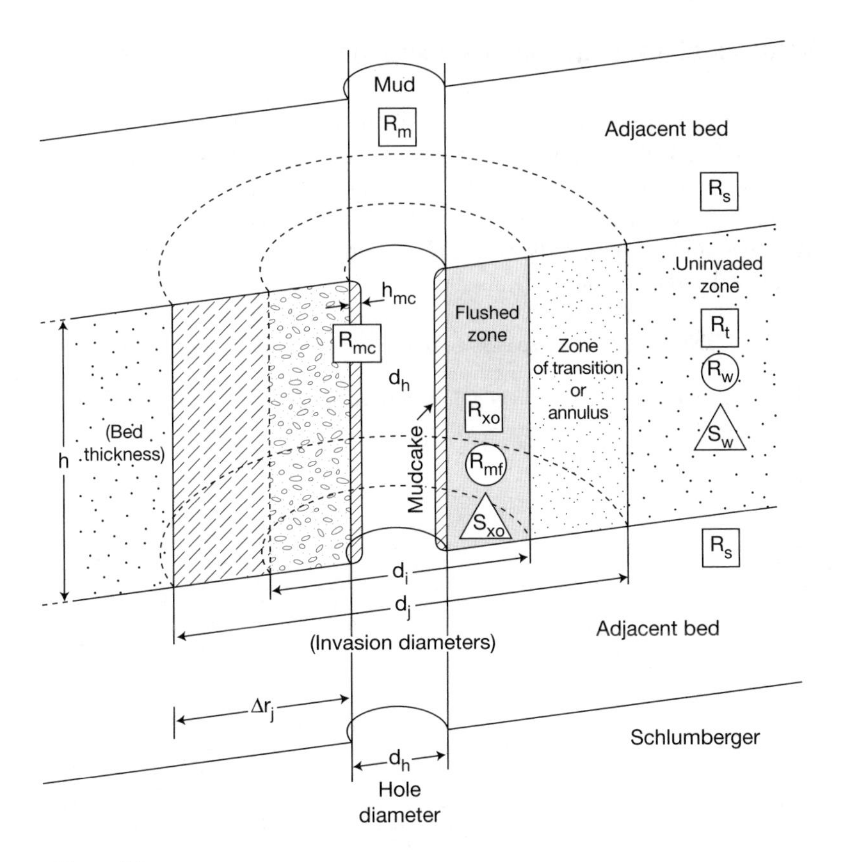

Figure 4.2

Volumes of investigation of logging tools: lateral view. Measuring the uninvaded zone cannot be done without accounting for the volumes closer to the borehole. By courtesy of Schlumberger.

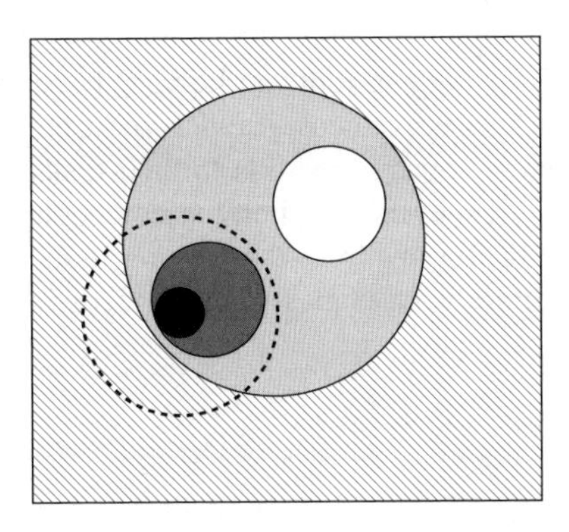

Figure 4.3

Volumes of investigation of logging tools: cased-hole configuration.

4.1.3 Volumes of investigation of logging tools: some numbers

When a measurement is performed at surface (scale, ruler, etc.), the object under investigation is clearly defined. This is not the case in well logging. Table 4.1 lists the numbers published for the volumes of investigation for most resistivity tools. The table is split between vertical resolution and depth of investigation [1]. It is clear that the formation characteristics can vastly differ (thin or thick beds, no-, shallow- or deep-invasion, variable formation features – fractures, karsts, pebbles, etc.).

Table 4.1 Vertical resolution and depth of investigation of logging tools

Logging tool	Vertical resolution (in)	Depth of investigation (in)
Imager electrode	0.2	12
Microlog normal	2 to 4	2 to 4
Micro-SFL	5	12
Short normal	12	12
Normale	90	60
Laterolog shallow	24	25 to 50
Laterolog deep	24	80
Induction 10 in.	1, 2, 4 ft (90%)	10 in. (50%, 0 mS/m)
Induction 20 in.	1, 2, 4 ft (90%)	20 in. (50%, 0 mS/m)
Induction 30 in.	1, 2, 4 ft (90%)	30 in. (50%, 0 mS/m)
Induction 60 in.	1, 2, 4 ft (90%)	60 in. (50%, 0 mS/m)
Induction 90 in.	1, 2, 4 ft (90%)	90 in. (50%, 0 mS/m)
Deep resistivity	8 ft (90%, < 10 ohm-m)	60 in. (50%, 0 mS/m)

These figures are approximate as vertical resolution and depth of investigation vary with bed thickness, invasion profiles and mud resistivity.

The volumes observed by a logging tool can be very large for resistivity tools. [3] They include the borehole, several successive beds, invaded zone, etc. Fig. 4.4 represents the signal contributions from different zones in relation to their distance to the logging tool. The zones contributing the most are in black. Smaller, but non negligible contributions are coming from volumes as far as 12 ft away from the logging tool. It is intuitive that extensive processing is needed to separate what comes from the beds in front of the tool, from the borehole and from beds further away.

3. This design results from the fact that the invaded zone occults the zones of interest (virgin zones with fluid saturations reflecting the state of the formation before drilling).

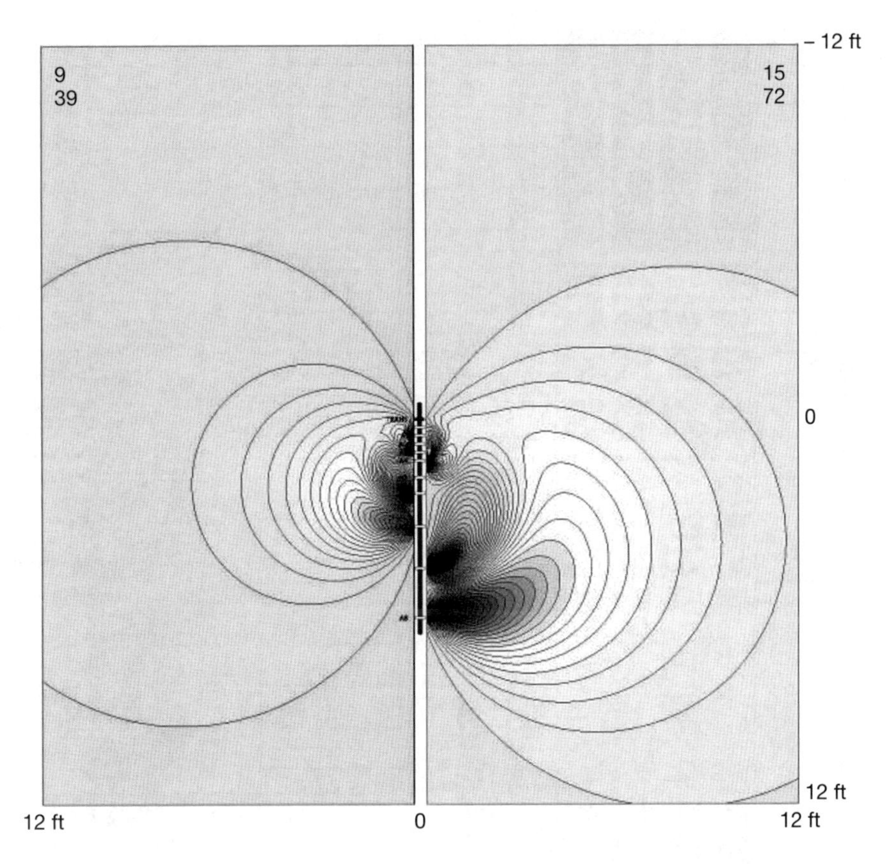

Figure 4.4

Volume of investigation of an AIT tool.
Note that the drawing represents a cylinder 24- ft tall and 12- ft in radius.
By courtesy of Schlumberger.

4.1.4 Turning high-resolution as an advantage

High-resolution is often linked to poor depth of investigation. Why are high-resolution logs run? In fact high-resolution sensors can be used to detect small rock features and better define thin beds. Fig. 4.5 is an example. The density-neutron displays a gas crossover pattern over a 9-ft interval from xx036 to xx027. The medium and deep resistivity curves confirm the pay interval. In the interval above, the resistivity is substantially lower, the gamma-ray reads higher and the density-neutron separation can be interpreted as a sandstone with a high shale content. The high-resolution conductivity curve, derived from a formation resistivity imaging tool, changes the picture. The zone is not a thick shaly sand formation but a sequence of clean hydrocarbon-bearing sands (below 200 millimhos, equivalent to 5 ohm-m) and shaly beds (above 1,000 millimhos, equivalent to 1 ohm-m). Twelve additional feet of net pay are detected, making the prospect an economical target. The resistivity images in Track 2 confirm

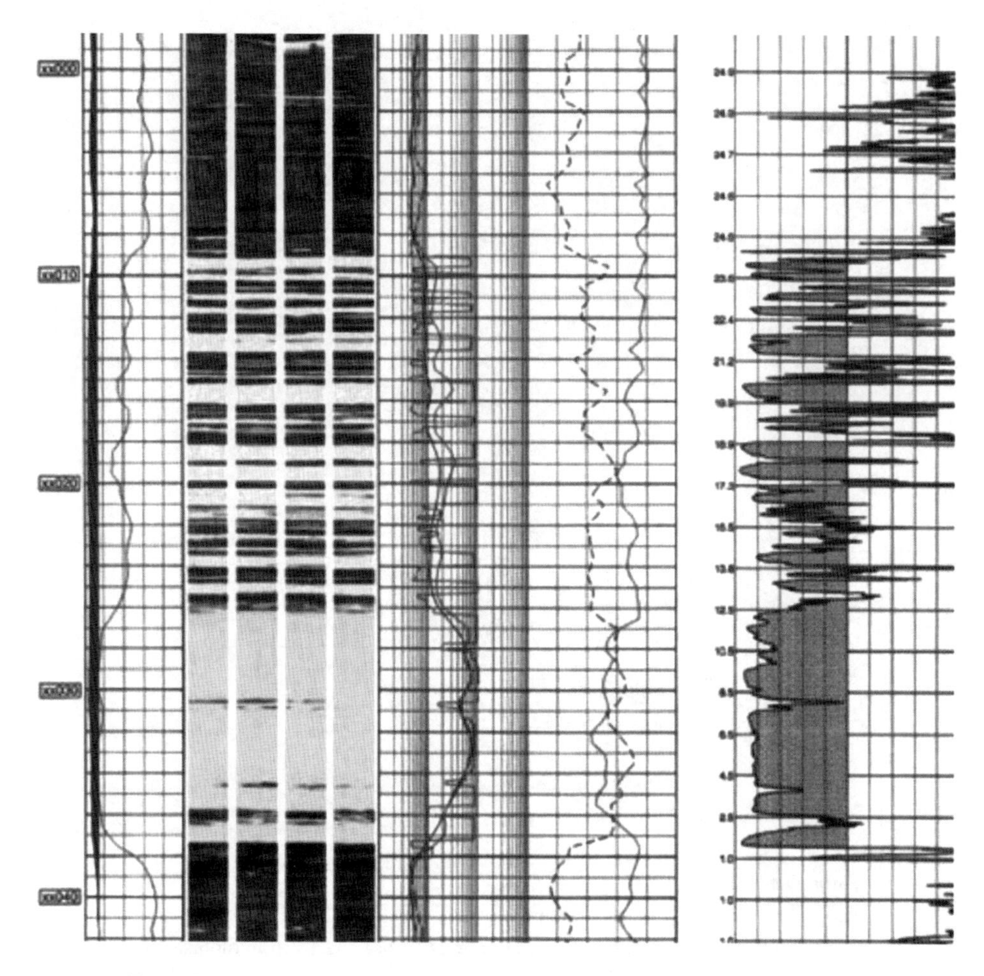

Figure 4.5

Sequence of thin beds.

The classical logs (medium and deep resistivities, as well as the density-neutron) do not “see” the thin beds, but the high-resolution imaging curves recognize them (the sequence of black and white segments is a succession of clean and shaly beds). The curve on the right track is built from the image information.

By courtesy of SPWLA.

the presence of thin laminations. Note that the conductivity curve does not give quantitative information, but enables a better description of the geometry of the rocks.

High-resolution information does not always yield higher hydrocarbon volumes. Lower hydrocarbon volumes may result from the use of high-resolution logs. In all cases, the combination of tools with different resolutions enables a more realistic description of the formations. It is to be stressed that high-resolution logs may not carry much useful information if the borehole condition is poor.

4.2 ENVIRONMENTAL EFFECTS

4.2.1 Borehole effects

Well logging, during or after drilling, implies the presence of the borehole. Practically, all logging tools receive signals from the borehole and are somewhat affected by it.

An extreme example: barite effect on the photoelectric factor

In the presence of barite, the logging sensor sees so much of the borehole that it cannot see the formation at all. In a fictitious (but not unrealistic) tool, the photoelectric sensor "sees" 97% of the formation and 3% of the borehole. It is reminded that linear volumetric equations cannot be written with P_e, but can be used with $U = P_e \times \rho_b$. A limestone formation is surveyed. $U_{formation} = 13.8$. The volumetric equation can be written:

$$U_{log} = 0.97 * U_{formation} + 0.03 * U_{borehole}$$

When there is no barite in the mud, $U_{borehole} = 0.2$. Then, $U_{log} = 13.4$, a 3% difference from the limestone value. This small shift can be corrected. When there is 25% barite in volume terms, $U_{borehole} = 1{,}091 \times 0.25 = 273$ and $U_{log} = 21.6$. The log value is 56% off the true value of the formation and cannot be corrected.

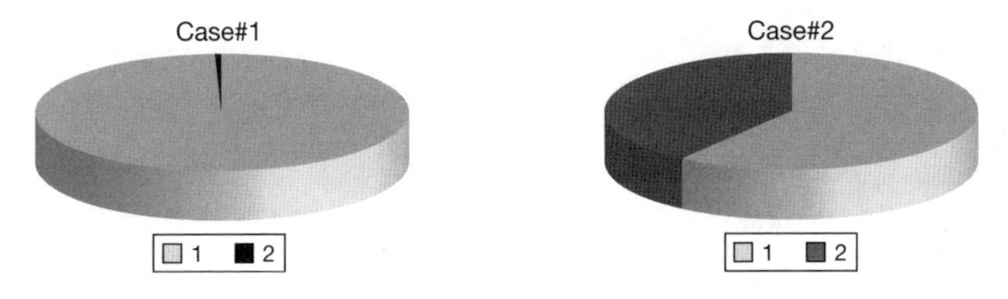

Figure 4.6

Consider an ideal non-existent logging tool that looks at 99.95% of the formation and 0.05% of the borehole in case #1, and at a more realistic case #2 where the split is 62%-38%. In case #1, the correction is possible even though $U_{borehole}$ is 20 times larger than $U_{formation}$. In case#2, the correction is not possible. Nevertheless the P_e and U curves can be used as excellent barite indicators.

Effect of the hole shape

The borehole may affect log readings in a different way. This happens when the shape of the hole precludes the correct positioning of the sensor. When the stress field on the rocks is not isotropic, the hole takes a lemon shape. The surfaces at the extremities of the long axis of the hole are rugose, while the extremities of the short axis are comparatively smooth. A sensor mounted on a pad reads much of the mud in the first case, but delivers reasonable readings in the second (Fig. 4.7).

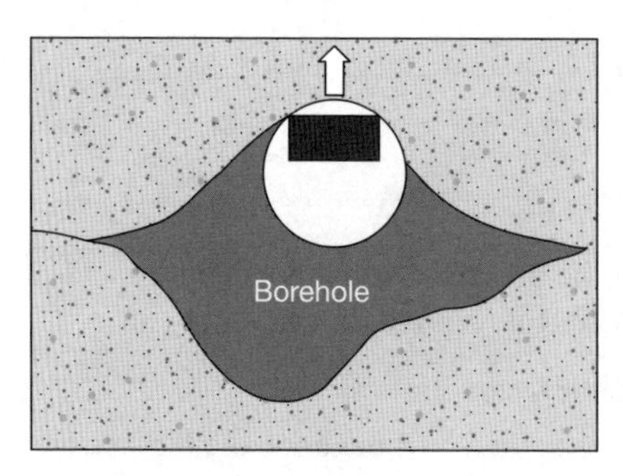

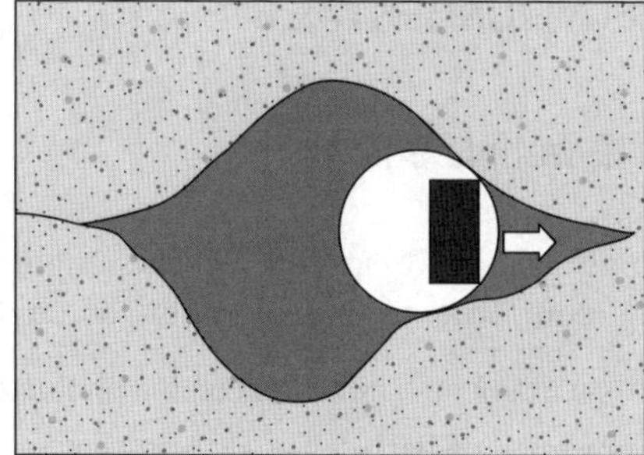

Figure 4.7

Effect of orientation of a pad tool. Left: Readings are correct. Right: Readings are affected by mud.

The borehole scanner tool

Considering the large impact of the borehole on logging tools, the design of a borehole scanner tool is a step forward. This tool has the following features:

- It reads the mud properties (resistivity, density, barite content).
- It describes the hole shape and size with its multiple calipers.
- It does not get stuck.
- It is compact.
- It does not affect the positioning of the other logging tools.

Unfortunately, this tool needs yet to be designed and commercialized.

4.2.2 Mud cake

Once the hole is drilled, the difference of pressure and fluid composition is followed by the buildup of a mud cake between the mud and the formation. The mud cake affects shallow investigating devices, and in particular the density log. An ingenuous idea was tried to compensate for mud cake effects. Two detectors at different distances from a nuclear source are used. This enables the observation of separate zones, a shallow one and a deeper one. The information is collected simultaneously (i.e., not in separate runs). This strategy is now commonly used for multi-spacing resistivities and is used to unravel the next challenge, tackling invasion.

4.2.3 Invasion

As the borehole remains open to borehole fluids and the mud pressure is higher than formation pressure, filtrate invades the virgin formation, with a resistivity R_t. The way to measure uninvaded formation resistivity is to design deep reading tools. But these tools also "see" the more shallow zones (and even the borehole). Therefore, shallow-, medium- and deep- investigating measurements are performed simultaneously to enable a reasonable solution of R_t.

4.2.4 Shoulder beds and formation dip

Deep reading logging devices, needed to observe beyond the invaded zone, tend to have low vertical resolution and see beds above and below the zones of interest. Large differences between real formation resistivity R_t and deep readings are observed when the beds are thin (less than 10 feet thick). These differences are aggravated by the presence of dip [2]. In Fig. 4.8, the modeled Rt is shown in tracks 2 to 4. The measured and modeled medium and deep induction are also shown. Larges differences between R_t and the measurements are observed when resistivity reads above 20 ohm-m. In zone D, the deep resistivity reads 20 ohm-m, that is only 25% of R_t (at 80 ohm-m).

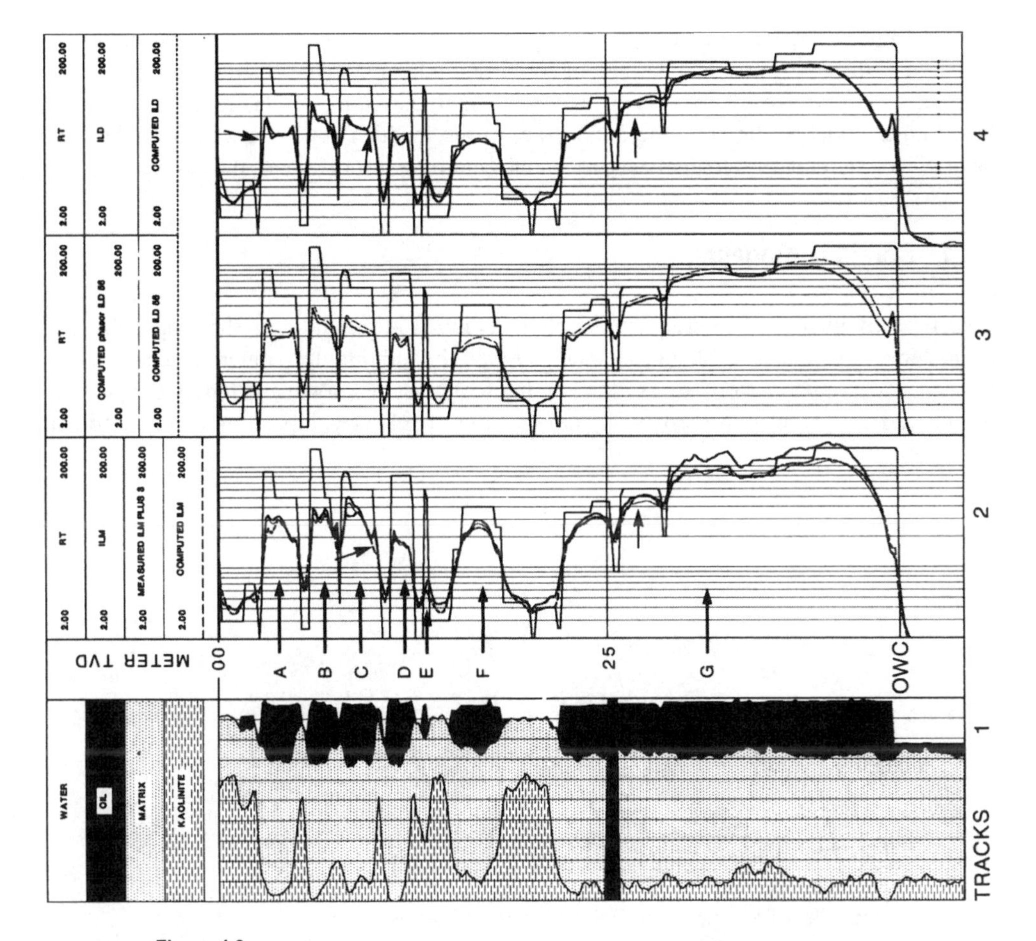

Figure 4.8

Resistivity readings in highly dipping formations. The dip is 56°. Tracks 2 to 4 display modeled and measured values for medium and deep induction logs.
By courtesy of SPWLA.

In dipping thin beds, it often happens that the shallowest curve (e.g., the 10-in curve from the Schlumberger AIT tool) is closest to R_t [3]. It is a paradox that the industry often uses the deep reading as the best estimate of R_t. In the case of thin dipping beds, selecting the deep reading corresponds to the largest error.

Considering the large errors observed when the logging tool axis is not perpendicular to the bedding of the formations, three-dimensional induction tools have been developed. With them, it is possible to measure dip and correct the resistivity curves for it [4].

4.3 MODELING THE ENVIRONMENT

Considering the large disparities between measured values and real values as a result of the presence of the environment, a number of correction schemes have been developed. All logging companies have characterized "environmental corrections." These corrections are based on a model of the borehole, mud cake, of the invaded zone and of the neighboring beds. This a step forward to get measured values more representative of the real values.

4.3.1 Borehole modeling

To efficiently manage the borehole correction, as much information on the shape, size and characteristics of the borehole needs to be gathered. In addition, the centering scheme (centralized, excentralized, stood-off) of the logging tool must be known and reported. Knowing the properties of the mud (resistivity, density and chemical composition) is also critical (Fig. 4.9).

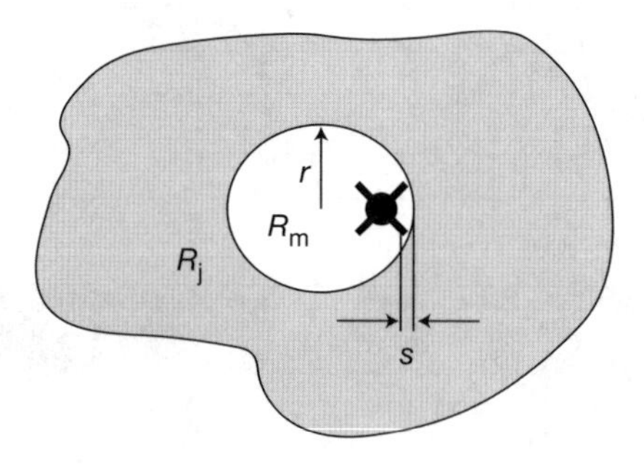

Figure 4.9

Modeling the borehole effect

A number of parameters depicting the hole are needed.

By courtesy of Schlumberger.

4.3.2 Mud cake modeling: spine and ribs

The mud cake effect is automatically corrected through a method called "spine-and-ribs." In absence of mud cake, only the formation density ρb is of concern. Handling mud cake requires the management of five parameters, ρ_{mc}, t_{mc}, Z_{mc}, ρ_b and Z_b as shown in Fig. 4.10.

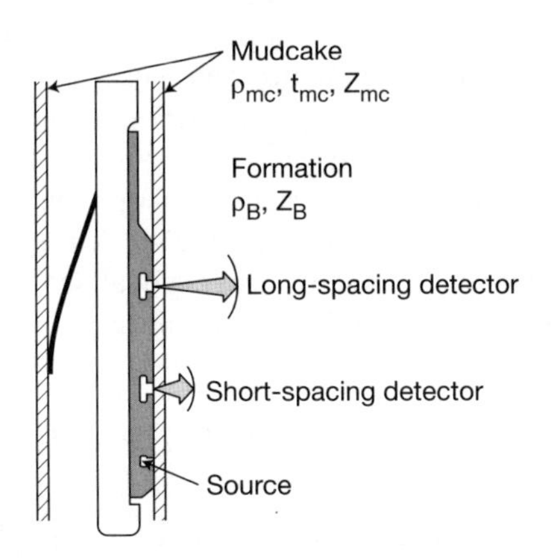

Figure 4.10

Density model leading to the spine-and-rib method.
By courtesy of D. Ellis and J. Singer.

4.3.3 Invasion modeling

In addition to measurements performed at different depths of investigation, correcting for invasion needs the selection of an invasion model. Fig. 4.11 displays three possible schemes. Many parameters are introduced by these models. In the case of the step profile, there are three unknowns, R_{xo}, R_t and r_i, radius of invasion. In the slope profile, there is an additional parameter. If an annulus profile is assumed, then five parameters, R_t, R_{xo}, R_{ann}, r_1 and r_2 are solved.

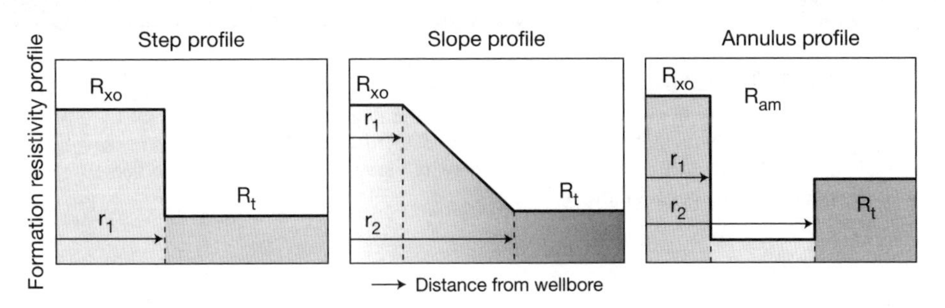

Figure 4.11

Possible invasion models.
By courtesy of Schlumberger.

4.4 LIMITATIONS OF ENVIRONMENTAL MODELING

Most environmental models are simple. The reality may not match with the model. For instance, the correction model for density (Fig. 4.10) assumes a smooth borehole surface. Fig. 4.12 shows two density measurements, one acquired while drilling, the other one after drilling, by wireline conveyance. The LWD measurement is strongly affected. The correction curve on the right displays large variations. The density curve itself shows large swings with an amplitude of 0.3 g/cm^3, making any quantitative use invalid. The caliper, run after drilling explains the issue. The hole diameter varies by one inch over a short distance. The density sub is unable to stay close to the formation as represented on Fig. 4.13. Why is the wireline density less affected? It is because the wireline sensors are mounted on a shorter pad that can track the difficult hole geometry. This is not all the time the case and some hole geometries may be more favorable to LWD than wireline.

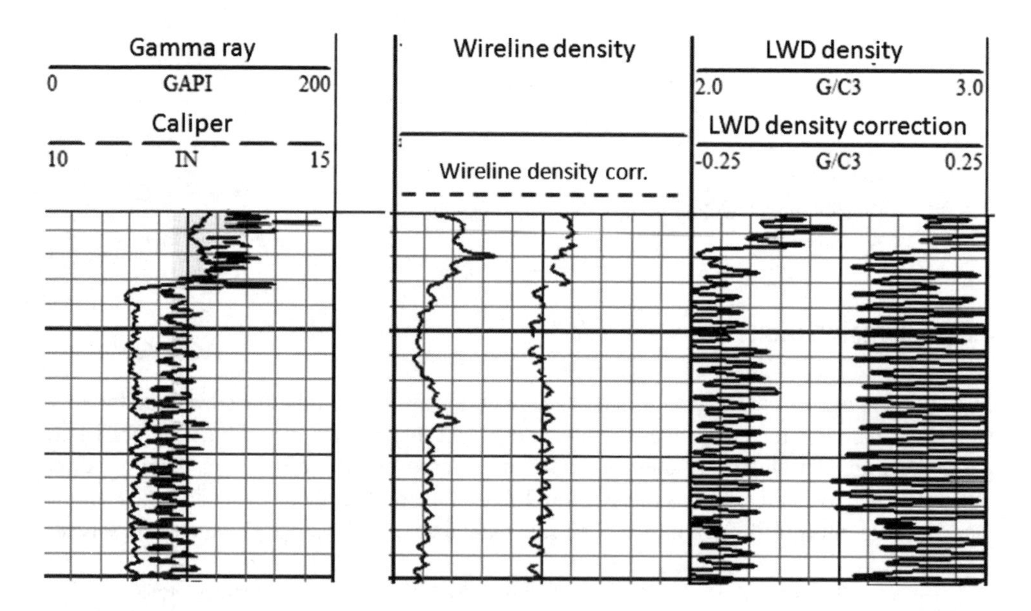

Figure 4.12

Instance where the borehole model used in the density algorithm does not fit with reality. Note that the wireline density is not affected. The caliper has been acquired after drilling. In track 3, the density curve is on the left and the density correction curve on the right.

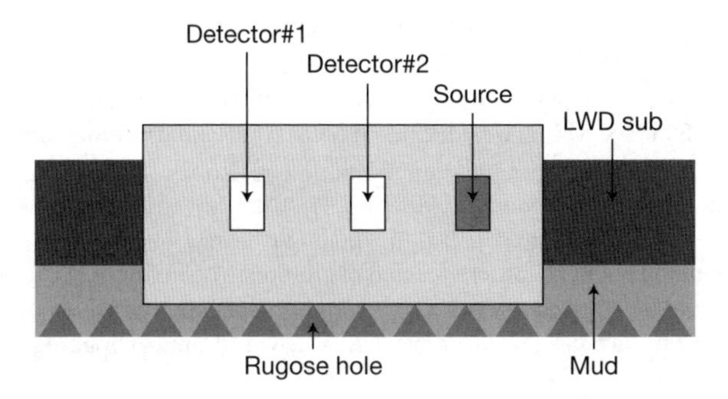

Figure 4.13

Reality does not fit with the model used to correct the density. There is a lack of synchronization between the shape of the hole and the spacing of the LWD tool.

4.5 SUMMARY

- Logging tools do not exclusively survey the zones of interest.
- Logging measurements are affected by the borehole, invasion and the neighbor beds in open holes.
- Logging measurements are affected by tubings and casing, cement, various types of production fluids in cased holes.
- Environment models are used and may correct the readings.
- Environment models cannot cope with all real situations and may not always represent all the conditions that can be encountered in reality.

REFERENCES

[1] Flaum, C., Theys, P., "Geometrical specifications of logging tools: the need for new standards," SPWLA 32nd annual logging symposium, Midland, 1991.

[2] Fylling, A., Spurlin, J., "Induction simulation, the log analyst's perspective," 11th European formation evaluation symposium, Oslo, 1988.

[3] Hartmann, R., personal communication.

[4] Wu, P., Barber, T., Homan, D., Wang, G., Johnson, C., Heliot, D., Kumar, A., Ruiz, E., Xu, W., Hayden, R., Jacobsen, S., "Determining formation dip from a fully triaxial induction tool," 51st annual logging symposium, Perth, 2010.

Additional readings on resistivity measurements and on the quest for values closer to R_t:

[5] Barber, T., "Introduction to the Phasor Dual Induction Tool," *J. Pet. Tech.*, 37, N° 10, pp. 1999-1706, 9-1985.

[6] Barber, T. and Rosthal, R., "Using a Multiarray induction tool to achieve logs with minimum environmental effects," paper SPE 22725, 66th SPE annual technical conference and exhibition, Dallas, 1991.

[7] Clark, B., Lüling, M.G., Jundt, J., Ross, M.O. and Best, D., "A dual depth resistivity measurement for formation evaluation while drilling," SPWLA annual logging symposium, 1988.

[8] Crary, S., Jacobsen, S., Rasmus, J., and Spaeth, R., "Effect of resistive invasion on resistivity logs," Paper SPE71708, SPE annual technical conference and exhibition, New Orleans, 2001.

[9] Béguin, P., *et al.*, "Recent progress on formation resistivity through casing," 41st SPWLA annual logging symposium, Dallas, 2000.

[10] Davies, D.H., *et al.*, "Azimuthal resistivity imaging: a new generation laterolog," paper SPE 24676, SPE annual technical conference and exhibition, Washington, D.C., 1992.

[11] Griffiths, R., Barber, T., and Faivre, O., "Optimal evaluation of formation resistivities using array induction and array laterolog tools," SPWLA 41st annual logging symposium, Dallas, 2000.

[12] Maurer, H.M., and Hunziker, J., "Early results of through-casing field tests," 41st SPWLA annual logging symposium, Dallas, 2000.

[13] Suau, J., Grimaldi, P., Poupon, A. and Souhaite, P., "The dual laterolog-Rxo tool," paper SPE 4018, SPE annual technical conference and exhibition, 1972.

[14] Smits, J.W., *et al.*, "High-resolution from a new laterolog with azimuthal imaging," paper SPE 30584, SPE annual technical conference and exhibition, Dallas, 1995.

[15] Smits, J.W., *et al.*, "Improved resistivity interpretation using a new array laterolog tool and associated inversion processing," paper SPE 49328, SPE annual technical conference and exhibition, 1998.

[16] Tanguy, D.R., "Induction well logging," U. S. Patent 3,067,383, 1962.

[17] Tanguy, D.R., "Methods and apparatus for investigating earth formations featuring simultaneous focused coil and electrode system measurements," U. S. Patent 3,329,889, 1967.

5

Measurements are imprecise and inaccurate

5.1 ELEMENTS OF REALITY

In the real world, acquisition of data is a one-time, unique, specific event. It happens once, and can never be repeated exactly again.

Data acquired by Engineer#1 is slightly different than the one acquired by Engineer#2, even if the same surface and down-hole equipment are used. If two different tools (with similar physics and design) are run, they would give different measurements. If two logging companies are put on the same well, they would acquire different records. Even if the same engineer uses the same equipment and surveys the same formation twice, the logs will be different. This is highlighted by comparing the main pass and the repeat pass.

All those measured values are different from the true value. The difference between the true value and the measured values, the uncertainty is what needs to be quantified, or, if it is not possible, estimated.

5.2 SOME DEFINITIONS

5.2.1 Academic definitions [4]

- **True value**: The value towards which the average of single results obtained by n laboratories tends, as n tends towards infinity.
- **Uncertainty of measurement**: Result of the evaluation aimed at characterizing the range within which the true value is estimated to lie, generally with a given likelihood.
- **Observational error**: It is the difference between the measured value and the true value. It is not a "mistake." Variability is an inherent part of things being measured. Systematic errors (or bias) and random errors are the main types of error.
- **Accuracy** of a measurement is the closeness of the measured value to the true value. Its magnitude depends on how systematic errors are controlled.

- **Precision** is the closeness of two measurements made with the same method. Its magnitude depends on how random errors are controlled.
- **Repeatability** is the difference of two measurements made under the same conditions, with the same equipment, the same engineer and the same environment.
- **Reproducibility** is the difference of two measurements made with the same method but possibly with different equipment and different personnel.

Fig. 5.1 is a graphical representation on how true and measured values relate. Fig. 5.2 shows that accuracy and precision are not linked.

1) Inaccurate and precise,
2) Imprecise and accurate, are possible states of data.

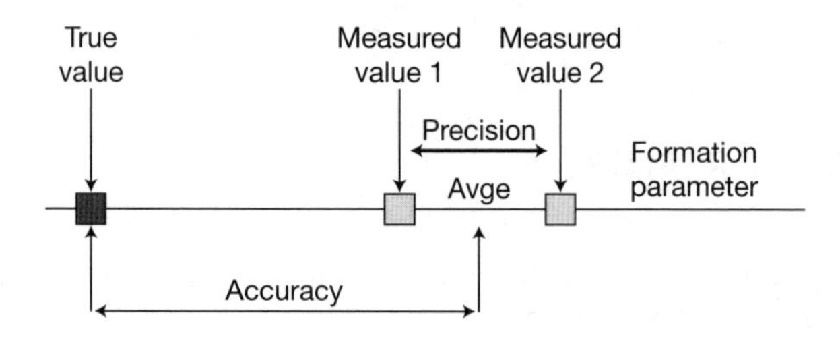

Figure 5.1

Graphical representation of accuracy and precision.

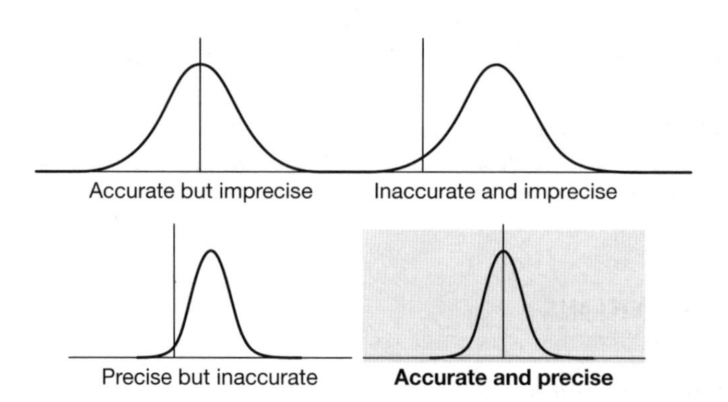

Figure 5.2

Precision and accuracy. The vertical line represents the true value. The thicker curve represents the locus of the possible measurements.

5.2.2 Practical considerations on errors

There is definitely some confusion associated with true and measured values, errors, accuracy and precision. Table 5.1 is an attempt to reduce this ambiguity. It links the different types of errors and the related control on them.

Table 5.1 Type of errors and methods to control them

Type of error	Method of control
Random	Precision control
Systematic	Accuracy control
Blunder	Data quality control
All errors except fudging	Uncertainty management
Nudging, fudging, hiding problems	Controlled by an integrity policy

The difference between the true value and the real value is an accumulation of errors. The quest for data quality is a hunt for these errors. When all the errors are unveiled, there is a better understanding of the distance from measured value to real value.

- Random errors are by definition statistical. They cannot be prevented. But they can be reasonably quantified by repeating the measurement. This is why it is possible to assign quantitative values to repeatability. The standard deviation of the various values collected over the same depth interval is an excellent evaluation.
- Systematic errors stay constant. When they are not identified, they cannot be corrected. But, once recognized, they can be completely compensated for. Accuracy, the closeness to the true value, can be also estimated by experience and by performing many measurements in different conditions, with different tools and different people. A good estimate of accuracy is reproducibility.
- Blunders, or involuntary human errors, can be harvested from a close scrutiny of the data set. Thorough reporting by the person performing the data acquisition can be helpful as it may highlight inconsistencies.
- Voluntary human errors, such as shifting data without reporting it or hiding circumstances of the acquisition job are difficult to track as there is an intentional desire to falsify reality. Nevertheless, with the advent of complicated data sets, it is difficult to create a coherent data set that has been tampered with. For instance, if calibration dates are edited, the calibration coefficients must follow some trends linked to the physics of the tool. It is almost more difficult to artificially create a credible calibration than performing the calibration itself. In any case, intentional errors can be reduced by the deployment of a strict "Integrity policy." [1]

5.2.3 Addition of errors

Errors are generally added in a quadratic way. For instance, if there are two errors δ_1 and δ_2, the total error is $\Delta\varepsilon$, such as:

$$\Delta\varepsilon^2 = \delta_1^2 + \delta_2^2$$

What does that mean? If M is the true value of the formation parameter, all measurements m_1, m_2, m_3, etc. would lay between: [2]

$$M - \Delta\varepsilon \text{ and } M + \Delta\varepsilon.$$

1. See Chapter 17.
2. In fact, a probabilistic approach would state that there is 68.3% chance that the measurements lie between $M - \Delta\varepsilon$ and $M + \Delta\varepsilon$, if data behaves in a Gaussian way [5].

In practical terms, as M cannot be known, a slightly parallel approach is used. Starting from measurements, it is possible to take the average of them, m_a, and claim with a certain level of confidence, that the true value M is between:

$$m_a - \Delta\varepsilon \text{ and } m_a + \Delta\varepsilon.$$

If only systematic error ε_s, and random error ε_r are accounted for, the measurements are averaged by m_a, and the true value M is such that:

$$m_a - \text{sqrt}(\varepsilon_s^2 + \varepsilon_r^2) < M < m_a + \text{sqrt}(\varepsilon_s^2 + \varepsilon_r^2)$$

5.3 WHAT AFFECTS PRECISION

Precision is the control of random errors. It varies with logging speed, sampling rates, filtering schemes and detector technology.

5.3.1 Logging speed and sampling rate

Precision depends on the length of time taken to perform the measurement. The more time, the better is the precision. In a well environment, it consists in spending more time in front of the same piece of rock. When converted in terms of logging speed or of rate of penetration, this means that the slower, the better. A reduction by a factor 2 of logging speed or ROP would be followed by an improvement of precision by a factor sqrt(2). Conversely a doubling would result in a deterioration of precision by sqrt(2).

Tables 5.2 and 5.3 Change of precision with logging speed or rate of penetration

Logging speed	Porosity precision	
ft/h	pu	pu
900	2/sqrt(2)	1.414
1,800	2	2.000
3,600	2sqrt(2)	2.828

ROP	Porosity precision	
ft/h	pu	pu
50	2/sqrt(2)	1.414
100	2	2.000
200	2sqrt(2)	2.828

5.3.2 Filtering

When the formations are thick and homogeneous, it is legitimate to average the values of several consecutive measurements. This average is more precise than individual measurements. If the formations are thin, then the average is not representative of the beds as it has not the vertical resolution to quantitatively assess individual beds. The well logging industry uses a vast choice of filters and signal processing scheme. Filtering is not a good practice or a bad practice. But its existence needs to be carefully documented.

5.3.3 Technology

Improved detector technology contributes to better precision. Two tools, run with the same logging speed, the same sampling rate and the same filtering scheme would have different precision. The one with higher counts would deliver better precision. Unfortunately, improved technology has often been used to increase logging speeds instead of improving precision.

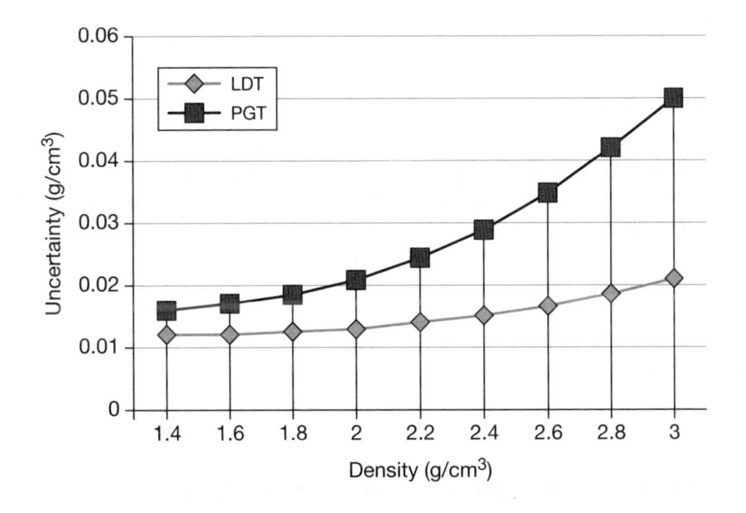

Figure 5.3

Diagram of repeatability PGT/LDT.

Fig. 5.3 represents the precision of two logging tools. The PGT was the first density tool to be commercialized. The LDT was deployed in the field in the early 1980s. Because of its improved detectors, the LDT has better precision. Note that the precision varies with the density that is measured. It is better for low densities than for high densities. Important variations from the nominal value in anhydrite ($\rho_{hanhydrite} = 2.98$ g/cm^3) are more acceptable than in salt ($\rho_{salt} = 2.04$ g/cm^3).

5.4 WHAT AFFECTS ACCURACY

Accuracy quantifies the difference between true value and measured value. As the true value is unknowable, it is more practical to use the second definition of accuracy, that is, the control of systematic errors. The potential systematic errors that may affect logging measurements are:

- Errors induced by the tool response derivation.
- Errors induced by the differences between logging tools.
- Errors introduced by the well environment.
- Errors introduced by the assumptions made about the environment.

5.4.1 Tool response

As seen in Chapter 3, logging tools do not directly measure the formations. Raw data is collected, and an algorithm between the raw data and useable data must be built. Different techniques are used [5] and introduce small errors. For instance, it is not possible to have experimental blocks for every possible value. Hence, some interpolation, accompanied with minor errors, is needed.

5.4.2 Tool calibration

Once the tool response has been defined with a number of prototype tools, the logging tools are deployed in large numbers. These tools cannot be exact clones of the original prototype series. Geometrical dimensions slightly differ, even if expressed in fractions of in. or of mm. Detectors cannot have the exact same efficiency. Logging sources (gamma rays, neutron sources) cannot be identical.

These differences are handled by calibrating the logging tools. This is mostly completed in the field, or in dedicated calibration centers such as the one calibrating pressure quartz gauges. Calibrations cannot be perfect and introduce systematic errors.

5.4.3 Environmental corrections and environmental models

The impact of the well environment has been described in Chapter 4. Two situations may take place:

- The environment is described by some auxiliary equipment or measurement. This is the case when a caliper is available or when the mud engineer makes a detailed description of the mud composition and characteristics. In these conditions, and if the proper environmental corrections have been characterized, it is possible to perform corrections that compensate the errors induced by the well environment. As the inputs to the corrections are also measurements, errors are propagated from these auxiliary measurements to the main ones.

- There is no quantitative information on the well environment. In that case no correction is applied. The experienced log analyst is well aware that this lack of knowledge introduces an uncertainty that cannot be quantified. He conducts his interpretation with this information in mind.

In most cases, the environmental corrections are quantified in narrow conditions, corresponding to a simplification of the real environment. The environmental model needs still to be reasonably close to reality. Otherwise, large inaccuracies occur.

5.4.4 Environmental effects

Table 5.4 lists the types of environmental effects and their conditions of occurrence.

Table 5.4 Environmental effects

Borehole size	Over nominal bit size Ledges and bridges	
Borehole shape	Borehole shape. Ovalization. Rugosity Lemon-shape	
Borehole trajectory	Apparent dip Dog legs Poor stabilizer or sensor application	
Mud composition	Oil base mud Mud solids Potassium in mud Barite	 Effect on density Effect on gamma ray Effect on density and P_e.
Mud distribution	Non-homogeneous distribution along TVD. Segregation. Invasion Invasion shape In highly deviated wells, invasion is not symmetrical around the well axis	
Temperature	May not follow a unique gradient. Varies with rocks thermal conductivity	

5.5 REPEATED MEASUREMENTS

Accuracy and precision as described in the previous sections can be estimated with modeling and experiments. Many assumptions are used in the process. There is an easier and more representative way to understand the variability of data: perform several measurements.

- In almost similar conditions, it consists in observing if there is much difference between main and repeated measurements. This approach is an instinctive trait with persons who spend time in laboratories. All medical tests (on blood and urine) are performed at least twice to verify consistency.
- Over the same formations, but with different logging tools. It is a test of reproducibility (see next paragraph).

5.5.1 Historical importance of repeated data acquisition

The first historical repetition of measurements was performed on the request of Dr Mekel, from Royal Dutch.

5.5.2 Repeated measurements as an epiphany

Obtaining two different curves for the same parameter is a depressing, but useful rite of passage. The newly initiated person quickly captures the fact that none of the two measurements is actually identical to the true value and that variability of measurements needs to be managed and controlled. Lack of consistency between information sources supposed to describe the same rock characteristics is often the opportunity of creating more exact models and putting energy, but also time, in leveling off the differences.

5.5.3 Repeat sections

Performing the same measurement over a limited depth interval has been a standard operating procedure for most data vendors. Unfortunately, this valuable information is often poorly used. The repeat section can be looked at on the graphical version of the deliverable. The main pass and repeat pass curves are sometimes displayed together with an area coding between them to highlight major discrepancies (Fig. 5.4). But, it is often difficult to find a digital record for this second pass. By comparing numerically the main and repeat pass values (some depth-matching may be necessary to compensate for an irregular movement of the logging tool), it is possible to quantify the precision of the log. This value is calculated in situ and is specific to the well and the logging tool.

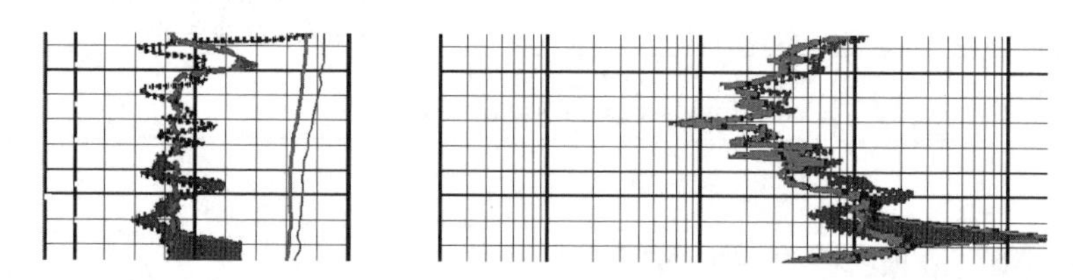

Figure 5.4

Example of repeat analysis: The main and repeat passes are presented together. Grey areas represent zones where the measurement does not repeat well.

5.5.4 Potential issues with repeat sections

To obtain quasi-identical values on several passes, two conditions are necessary:

- The same rocks (or casing) are observed during the two passes. This condition is not met if the measurement is only scanning a limited fraction of the formation. Sensors mounted on a pad see less than 25% of the hole circumference. As the tool rotates in the hole, it is possible that it will not follow exactly the same path on the two passes. The issue relates to azimuthal repeatability.

- The formation characteristics should not change with time. In logging while drilling, this condition is not met as drilling is a dynamic process. Resistivity changes as the result of invasion. Repeat passes in LWD require a different interpretation.

Optimal use of repeat sections is explained in reference [6].

5.5.5 Simultaneous measurements

Considering the inherent lack of focusing of sensors on the volumes of interest, data vendors have developed tools with many simultaneous measurements. The analysis of multi-sensor delivery is similar to repeat pass analysis. As the sensors have different volumes of investigation, the readings correctly display different values. But, there should be some coherence between them. The relative pattern of the curves enables a control of quality of the data.

This approach applies well to LWD. There could be only one sensor, with limited volume of investigation. But, because the drill string rotates, many measurements are performed over rocks of similar properties (unless there is poor angular continuity). Fig. 5.5 represents the density image derived from the observation of 16 azimuthal segments.

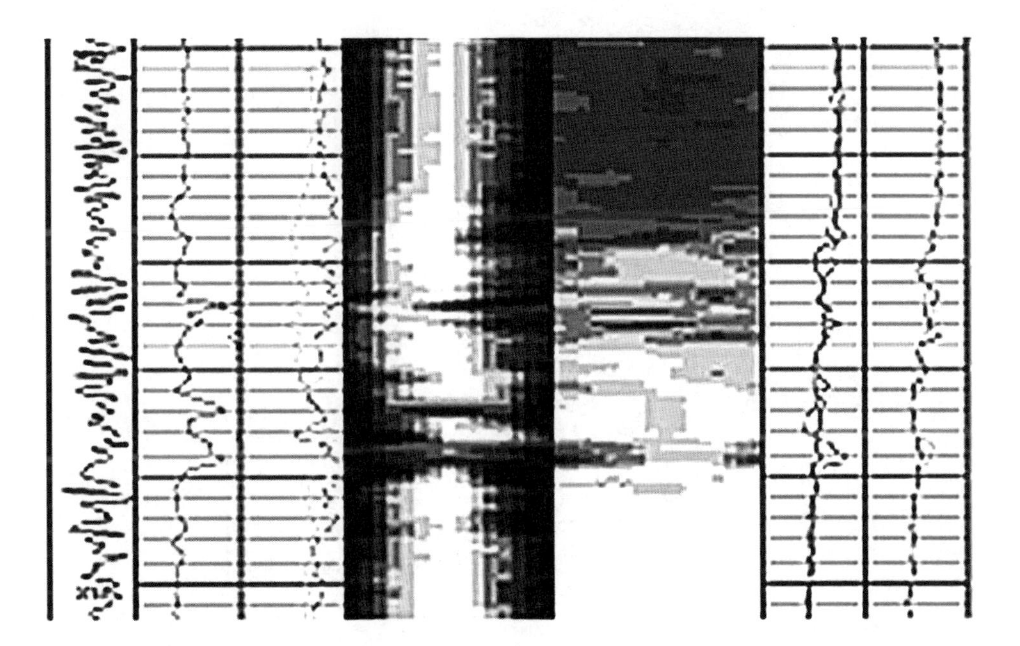

Figure 5.5

The density is measured azimuthally. As the LWD sub rotates, the formation is measured over 16 segments.

Some logging tools have several detectors that measure similar volumes. The design of several detectors compensates the effect of poor statistics. These detectors should read about the same values. Comparison of the raw data acquired by multiple detectors or multiple receiver coils is the basis of the quality control (QC) curves (see Chapter 16).

5.6 REPRODUCIBILITY

5.6.1 Historical perspective

Different measurements of the same formation parameter (resistivity) with different tool configurations were common (Fig. 5.6) in early logging history. They yielded different curves. The Kansas type sonde delivers resistivities higher than the Oklahoma type sonde [1].

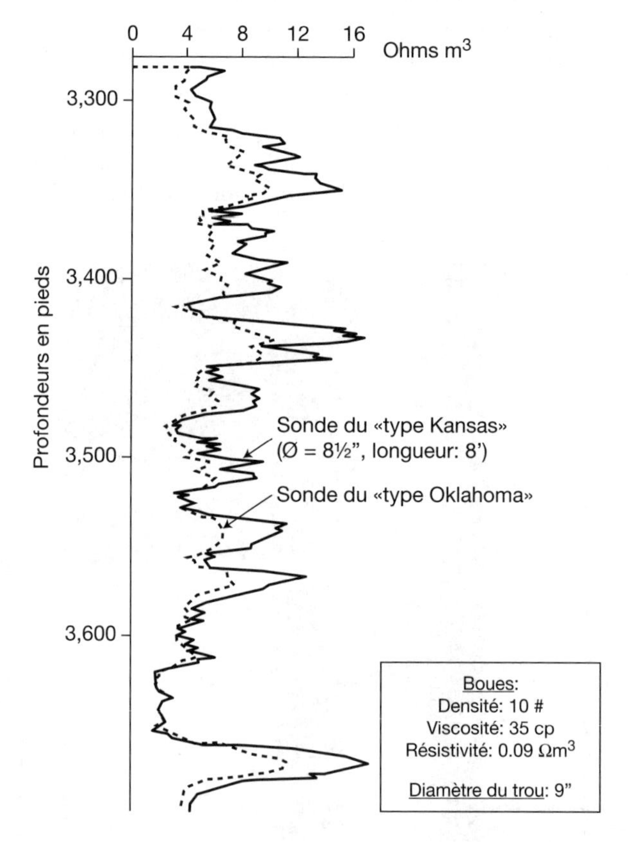

Figure 5.6

Different curves result from different electrode configurations. The auxiliary information of this early log is in French. The resistivity unit, Ωm^3, is not correct.
By courtesy of MIT Press.

By running two tools that give curves that are substantially different, it is easy to grasp the fact that none gives true resistivity and that additional efforts are required to understand the real characteristics of the formation.

5.6.2 Reproducibility tests

Large oil companies have run extensive reproducibility tests in the 1970s. As many as five different (but of the same type) sets of equipment were run over the same well intervals. For most tests, the results have remained confidential.

In the mid-1980s, a consortium of data vendors and oil companies was put together to drill new wells at the Conoco Bore Hole Test Facility in Newkirk, Oklahoma, expressly to compare LWD to wireline tools and analyze the recently introduced Nuclear Magnetic Resonance logs. Five LWD companies run their tools in the test facility. This organization produced a considerable wealth of information. Unfortunately, the data set, huge as predicted, was not fully analyzed. In addition, the wells did not produce!

Three observations are extracted from the test reports [2] and [3].

- In in-gauge holes, the LWD density had a tendency to read higher than the wireline density (Table 5.5).
- In washed holes (1-in washout), the LWD density has a tendency to read lower (Fig. 5.7).
- The epithermal neutron wireline tool has less statistical variation (or better repeatability) than the thermal tool (Fig. 5.8).

Table 5.5 Differences observed between densities run on MWD and on wireline for in-gauge holes. The LWD density is always higher than the wireline density, and this regardless of lithology. σ is the standard deviation. cc is the correlation coefficient.

	Limestone		Sandstone		All	lithologies	cc
	Δ	σ	Δ	σ	Δ	σ	
Company X	– 0.040	0.037	– 0.033	0.027	– 0.030	0.032	0.96
Company Y	– 0.022	0.028	– 0.034	0.050	– 0.011	0.035	0.95
Company Z	– 0.050	0.041	– 0.158	0.065	– 0.056	0.077	0.85
Company Z (Un-stabilized)	– 0.015	0.073	– 0.037	0.071	– 0.015	0.092	0.72

By courtesy of SPE.

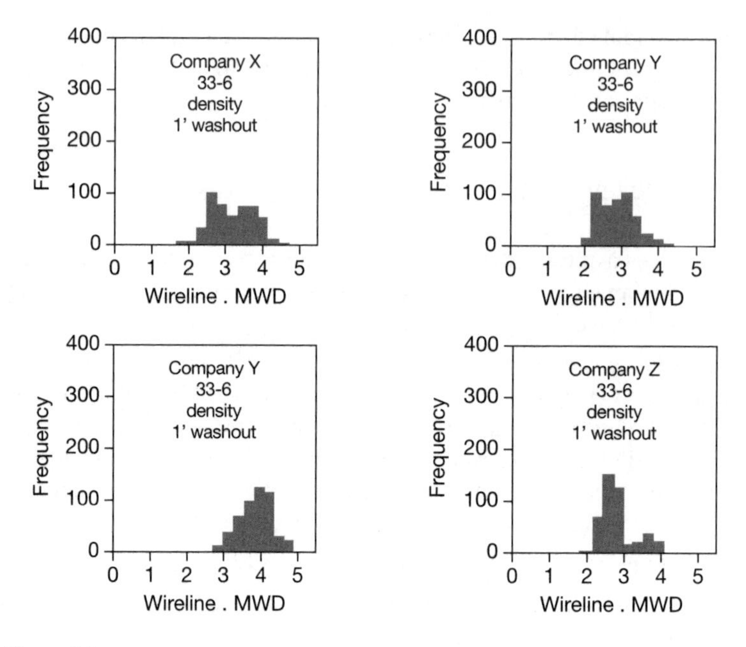

Figure 5.7

Differencces observed between densities run on MWD and on wireline in washout holes. The LWD density is mostly lower than the wireline density regardless of lithology.

By courtesy of SPE.

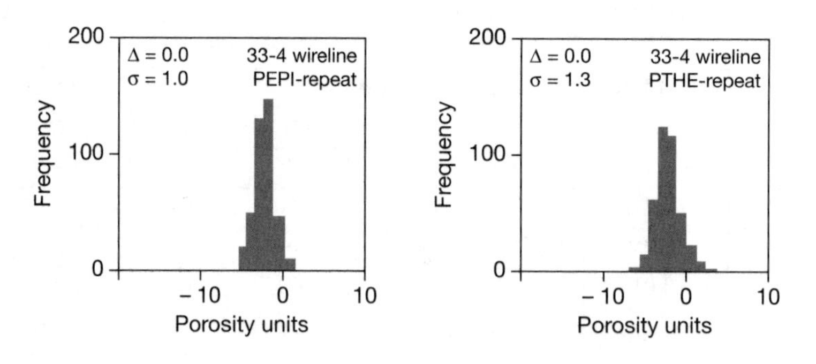

Figure 5.8

Differences observed between main pass and repeat pass for an epithermal (left) and a thermal (right) neutron porosity tool.

The standard deviation is larger for the thermal tool.

By courtesy of SPE.

5.6.3 Recent reproducibility studies

The recent trend, in opposite to the previous search for the understanding of data of different origins, is to avoid the confusion created by the presence of multiple values over the same formation. Some oil companies forbid the logging engineer to run a repeat section, while such a run is an invaluable complement of information. Anomalous readings are seldom surveyed again. In some extreme cases real-time LWD data goes straight to the oil company data base, and prevents any additional data, e.g., memory data to be used, even though memory data is more traceable and more robust.

In addition, not a single recent specification document has included the reproducibility tests performed by the data vendors.

5.7 WHAT MAY GO WRONG WITH A MEASUREMENT

The neutron porosity, φ_n, is used to exemplify what may distance the measurement from the real value. The φ_n measurement gives 27.34 pu. [3] This seems very well defined. In fact the following questions must be answered before this value is used:

- Is φ_n calibrated?
- Is φ_n corrected?
- Is this correction correct?
- What volume is investigated?

The value 27.34 pu could be:

- 23.14 if one of the system parameters (lithology) is set to a different value.
- 21.84 if the borehole correction uses a caliper value 1-in larger.
- 28.34 if it was raining during an outdoor calibration.
- 15.67 if there is a 8-ft depth offset and the sensor reads the tight streak below.
- 35 pu if the tool is run without a bow spring in a hole larger than 8 in.

5.8 SUMMARY

- All measurements are different from the true value.
- Errors affect measurements.
- They are different types of errors: random, systematic and human.
- Accuracy and precision can be analyzed to derive a quantitative value of the uncertainty.
- Repeating passes and performing logs with different tools bring useful information on the variability of the measurements.

3. pu means porosity unit.

REFERENCES

[1] Bowker, G., *Science on the run, information management and industrial geophysics at Schlumberger, 1920-1940*, The MIT Press, 1994.

[2] Hutchinson, M.W., "Measurements while and after drilling by multiple service companies through upper carboniferous formation at a borehole test facility, Kay county, Oklahoma," paper IADC/SPE 19969, Houston, 1990.

[3] Hutchinson, M.W., "Comparison of MWD, wireline and core data from a borehole test facility," paper SPE 22735, 66th annual convention, Dallas, 1991.

[4] International Organization for Standardization, *Quality assurance requirements for measuring equipment ISO 10012,* Genève, 1992.

[5] Theys, P., *Log data acquisition and quality control*, Éditions Technip.

[6] Theys, P., "A serious look at repeat sections," 35th SPWLA annual logging symposium, Dallas, 1994.

6

How measurements can suffer from human bias

For a successful technology, reality must take precedence over public relations, for Nature cannot be fooled.
Richard Feynman

6.1 INTRODUCTION

The limitations of physics imposed to the logging tools and measurements cannot be changed by the oil industry. Conversely, the human aspects of logging can be considerably influenced and corrected by the companies involved.

Well data has been reported to be insensitive to human bias in reference [3].

Two technological innovations were important because both of them generated objective, non-interpretive data:

Sensors were sent down wells, at the end of electrical cables, to measure rock properties. Known as wireline logs, they were run on virtually every well and are a rich source of data.

Sound waves generated at the surface and reflected back from subsurface rock layers produced cross sections similar to sonograms used in physician's offices.

As a confirmation, reference [5] affirms:

Sometimes logging sensors may deliver inferior measurements, but never do they intentionally lie.

Even though sensors do not lie, well data can be biased. This started early in the history of logging, as mentioned in reference [2]:

One contract driller offered to bribe two Schlumberger engineers to file a report that oil could not be found, since a successful strike would leave him without a job.

6.2 BIASED DATA

6.2.1 Examples

The following examples are all real. Details are not given to protect people and corporations.

- The while-drilling depth is adjusted to the driller's tally book depth to please him, even though the sequence of drillpipes put in the hole may have been misreported.
- The wireline depth is matched to the driller depth though they are measured in two different ways.
- The transfer of directional data from the MWD engineer to the directional driller is "manual." This opens the possibility of editing out some embarrassing doglegs that would nullify the bonus of the directional driller.
- A geologist has financial objectives based on porosity: the higher the porosity, the larger the bonus. The selection of the data acquisition company is based on who has measured porosities reading (inaccurately) high.
- A technical team takes months to evaluate some prospects to see the results increased arbitrarily by the manager by as much as 50%.
- The spikes on a logging-while-drilling resistivity curve are edited out because they do not look good. Unfortunately, the oil company is precisely expecting the spikes, associated with low porosity tight streaks.
- The curves look nicely smooth because a massive filtering scheme is applied to the data.
- Two gamma ray tools read differently over the same interval. There could be some good reasons like a change of composition of the mud or a possible enlargement of the hole. But, in order to minimize questions from an inquisitive data user, one of the gamma ray measurements is shifted to read values almost identical to the first one.

6.2.2 Where can data bias be introduced?

Raw data

Detectors and electronic circuitry have no knowledge of economics and profits. They provide unbiased data.

Calibrated data

Because master calibration coefficients (gains and offsets) are not generally accessible by an engineer or data user, calibrated data is mostly unbiased. But, calibration records can be tampered with. Another common example of nudging still happens: the change of caliper offset at the casing shoe.

Main curves

The main curves are presented after varying levels of processing. They can be affected by human intervention. Some examples are explained in later sections.

Control of the environmental corrections

In most cases, the field engineer has the option of enabling some corrections and disabling others. The effect on the processed curves can be very large as shown in Table 6.1. One engineer can arrive to a porosity of 25 pu by switching the temperature and pressure corrections "on" and the standoff correction "off." The second engineer can get a porosity of 18 pu by switching the temperature and pressure corrections "off" and the standoff correction "on."

Table 6.1 Obtaining different porosities from the same original raw data

All values in porosity units (pu)	Amount of correction	Case#1	Case#2
Original value		20	20
Temperature	4	Y	N
Borehole salinity	1	Y	N
Standoff	– 2	N	Y
Total correction		5	– 2
Corrected value (shown on the print)		25	18

Interpreted results

Log interpretation provides an additional layer of bias. Oil companies are indeed more pleased by good news, even if this information is not completely in line with reality.

Interpretation provides many opportunities to slant data in a convenient direction. The cement bond log is taken as an example. The sonic amplitude curve is shown on a cementation log. This information cannot be biased. But, because the amplitude needs to be converted for the right casing size and casing weight, logging companies have introduce a transform enabling the presentation of a bond index. The transform can be modified by the field engineer so that the final curve shows a more optimistic index, which means a better bond.

6.3 ARE DATA ACQUISITION COMPANIES SERVICE COMPANIES?

Bias on data is worsened by the fact that the oil industry considers that the data acquisition companies are service companies. Are they really?

6.3.1 Attributes of a service activity

"Service," as a deliverable is a customer-oriented "result." This result is produced when an organization performs activities that are oriented towards meeting customer expectations. Services can be described in terms of their main attributes [6].

1) Intangibility
   Service cannot be seen, handled, smelled, etc. There is no need for storage. Because it is difficult to conceptualize, "service" marketing requires creative visualizations to effectively make the intangible more concrete.

2) Perishability
   Service is perishable. Unsold service time is "lost," that is, it cannot be regained. Other service examples are airplane seats (once the plane departs, those empty seats cannot be sold), and theatre seats (sales end some minutes before the show).
3) Lack of transportability
   Service must be consumed at the point of production.
4) Lack of homogeneity
   Service is typically modified for each user or each new situation. It is often highly customized. Mass production of services is very difficult. This can be seen as a problem of inconsistent quality. Both inputs and outputs to the processes involved providing service are highly variable, as are the relationships between these processes, making it difficult to maintain consistent quality.
5) Labor intensity
6) Lack of consistency of processes
   Service usually involves considerable human activity, rather than precisely determined process. The human factor is often the key success factor in service industries.
7) Buyer involvement
   Most service provision requires a high degree of interaction between client and provider.

Only item 5 fits with the description of the activity of the logging companies.

- Logs are very tangible. Before 1980, they even smelled ammonia.
- The industry does everything so that log records do not perish. Archival is a definite concern for oil companies.
- Logs can be moved around worldwide in fractions of seconds.
- Data products need to be consistent and stable to be usable.
- As will be seen later, minimum buyer involvement is necessary to preserve the integrity of the data.

In conclusion, well logging companies do not have the attributes of the service companies. A better name is "data vendor."

6.3.2 Zero defect and zero pain

Another test to find out if an activity is product- or service- oriented is to check if it creates total client satisfaction, that is, zero pain. [1, 2]

- Service hands, as logging engineers are sometimes called, are bringing their share of pain to oil companies. They need to be flown on helicopters, fed and sheltered on offshore rigs.
- Logging crews tamper with the well for what always seems to be exaggerated lengths of time.

1. The quality goal of manufacturing a product is zero defect.
2. Jacques Horkowitz calls it zero hassle [4].

- Calibrations and justified repeat surveys of a logging interval take time. This is often seen as lost time or non productive time, NPT, while it is the only way to verify that the data is valid.
- Logging tools may require auxiliary equipment that is disliked by the drilling crew. For instance, the LWD density sub requires a stabilizer. The induction and neutron wireline tools require ex-centralizers.

According to the zero-pain criterion, logging companies are not service companies.

6.3.3 Bringing bad news, an important and difficult duty of the logging company

More importantly and even more painful for the oil companies, data companies sometimes bring bad news. In many instances, accurate information does not over-joy the oil and gas company [7].

- On about five out of six wildcat exploration wells, the logging engineer delivers data that says the well to be a duster and should not have been drilled.
- The caliper indicates that the hole is inches above bit size when the drilling engineer would have hoped for a nice, circular, in-gauge hole.
- The cement bond log indicates that there is no hydraulic integrity in spite of the large quantity of cement that has been pumped.
- The porosity logs show that the target reservoir is not of the expected quality.
- A directional survey reports that there are many doglegs in a supposedly perfectly curved well.
- Another directional survey confirms that the pay zone has been drilled in the neighbor's lease.
- The auxiliary sensors indicate that multiple and complex corrections will need to be performed before the main logs reach the correct level of usability.

In the job of collecting information, technical integrity is more important than bringing pleasing news. **It is still better to deliver accurate bad news than inaccurate good news.**

- Early collection of bad news enables the oil and gas company to take remedial action.
- A cement squeeze can be planned.
- Early plugging and abandonment of a dry well reduces the waste of money on testing.
- A sinuous well can be fixed before a casing gets stuck in it.
- Extensive correction of information after collection of the relevant environmental parameters makes a data set fit for quantitative evaluation.
- Early recognition of inferior data warns the user not to take decisions based on this data.

In all cases, the appraisal of the performance of the logging engineer cannot be based on his ability to get logs that are in pleasant agreement with the oil and gas company expectations. Candor and timeliness in delivering accurate information are more important.

6.3.4 Historical shift from product quality to service quality

In the beginning of the 20th century, the exploration methods were crude. Drilling, only on land, was inexpensive. Emphasis on data was high. Getting usable information came ahead of the time required to collect this data. A field engineer in the 1930s reports [2]:

We had the opportunity to work at a drill site where we were able to change the salinity of the drilling water. The results obtained fully confirmed the electrochemical theory. The research site had three wells, two drilled using sea water to flush them out, and the third using its own water.

Operating logging time was not an issue as late as the 1950s. Logging tools failed far more often than today and field engineers would spend rig time repairing them at the well site. Information was deemed critical and logging was pursued for days until a complete data set was acquired.

The importance of delivery (process and time) and of service quality evolved from perforating activities. In the late 1940s, perforating charges were bought from the same suppliers but used by different companies. In the highly competitive environment of the Lake Maracaibo, Venezuela, the companies strived to differentiate themselves and did it through optimization of operating time. Oil companies started using chronometers while field engineers performed the job. The product, "a hole in the casing," was deemed to be a commodity. It is only years later that perforating depth, skin damage and flow efficiency of the perforation were quantified, assessed and are now considered an integral part of quality.

Delivery time had the great advantage to be so simple it can be understood by all: the shorter, the better. It became a major aspect of quality. It is still a major component of the evaluation of suppliers, even though the products of the oil industry have become much more complicated, sophisticated and important.

6.4 OVEREMPHASIS ON SERVICE QUALITY HAS DETRIMENTAL EFFECTS ON OIL COMPANIES

6.4.1 Bias on reporting problems

This phenomenon is not unique to the oil industry and to suppliers. Trying to please customers or managers is a common human temptation. There is a natural reluctance to bring "bad news." There can be a financial bias when bonuses or incentives are involved. Expectations become a magnet and, as a consequence, reality is ignored. The supplier seldom reveals that a process has failed, especially when it is complicated and hard to understand or assess. The failure may not be detected before several years.

Product manufacturers, drillers for a well, acquisition engineers for data, interpreters for processed data, have an intimate knowledge of the quality of their products. Still, they are unlikely to spontaneously report a problem if it not obvious to the customer or cannot be easily explained.

Worse still, the overworked supplier may perform a smooth job and supply an excellent product. Because the product is not in line with the expectations of the customer, he may be tempted to make the product match this expectation, so that subsequent lengthy discussions are avoided. Most managers in the data acquisition companies repeatedly ask the field engineers to bring only satisfaction to the customers. Field engineers complain that every time they log a dry well, the customer thinks that they have performed a bad job and that the data product has flaws.

6.4.2 Selecting the most pleasant information

There are also many instances when decisions are taken on the basis of customer or management expectations instead of scientific or logical reasons. Sometimes, a product with several options is delivered. Most of the time, the selected option is the one that fits what is expected, instead of the one that makes the most sense in technical and scientific terms.

Formation resistivity is an example. Suppliers deliver resistivity curves with different resolutions. Stringent rules dictated by the logging tool design and laws of physics, control the validity of these curves. Nevertheless, the final curve is often selected because it fits an a priori expectation of formation resistivity.

Another example of the perverse effect of expectations is when several products relate to the same reservoir characteristics. Core descriptions and analyses from one supplier and interpreted data from another often coexist. Because the core, by its physical presence, tends to be trusted more than any other product, it is not rare that the interpreter develops a model that would enable a tight match between interpreted data and core description. Many interpreters have added trace minerals to their formation models to enable such a match while, in reality, discrepancies can be explained by the difference in volumes of investigation and the uncertainties in the measurements. The acquisition and interpretation chains are so complicated, that there is always a step in the process that can be used to adjust the data in the expected direction, whether it is a correction, a filter or the appropriate selection of an interpretation parameter.

6.4.3 Oil companies may destroy the ability to get correct data

Tools and measurement designers are aware of the borehole effects. Often, the only way to compensate for the borehole is to assemble specific positioning devices on the logging tools. These additions are deeply disliked by the oil company at the wellsite. It may happen that the field engineer is forced not to use the correct positioning devices because the oil company decides so. In most cases, the resulting data will be considerably degraded in quality.

6.4.4 Satisfying several oil companies may be ultimately impossible

Aiming at overall satisfaction is a difficult target when several partners are involved. The diversity of expectations makes delivery difficult to the supplier. It also happens that the same product is used by organizations with conflicting objectives.

For instance, a company geared towards exploration would like to see optimistic reserves so that it can "sell" the prospect in better terms. The companies specialized in production like to see pessimistic reserves so that they can demonstrate how well they exploit a mediocre prospect.

With databases, the industry enters a nightmarish domain. Companies mergers have been a definitive trait of the industry. The task of the data administrator in charge of organizing a single database is not enviable. Discussions with companies who have just merged indicate major difficulties to consolidate expectations and requirements. The culture of some companies is to expect every bit of data, even the most esoteric and arcane, to be delivered. In some other companies, it is radically the opposite, and the most simple and condense data sets are expected.

Another difficulty appears when a partnership includes an operator, singly interfacing with the suppliers, while the other partners have little interaction with the same supplier. These companies feel frustrated as their expectations are neither expressed nor met.

6.4.5 Satisfying some requests from an oil company may be illegal

When an oil company representative requests that a data acquisition company changes the aspect of data, there are some legal implications. A well record may be a government requirement linked to safety. Delivering this record with ambiguous or misrepresented information is criminal. An instance of that is to show a sonic-derived cementation indcx that does not actually reflect what the tool is reading, through the use of some intermediate processing parameters. It is also unethical to destroy directional data that shows that the well has penetrated the neighboring block.

6.5 LONG TERM USE OF DATA

In conclusion, most oilfield suppliers are actually delivering products, not services: pipelines, data, drilled wells. These products have a specific and common attribute: They have a very long life. They are often used longer than any other product in any other industry. Some early 20th century core data is still in use today. The first logs, run more than 80 years ago, can still be looked at, but the people who collected the information are long gone! [3]

6.5.1 An example of a century-old trouble-maker

The life of a well may exceed a century. Even though a well is no longer produced, it is still in the ground and may create chaos. Some years ago, a town in Central Kansas was riddled with explosions whose origin was mysterious [1]. Natural gas burst from the ground in several places. Two people were killed in one of the explosions. Thanks to 1960s geological

3. See more details on the first loggers in Chapter 17.

records, it was found that the gas had leaked from brine wells drilled in the early 1900s. The gas eruptions coincided with a leak of natural gas through a shallow hole in the casing of a wellbore drilled into a solution cavern in a salt bed of a gas storage field a few miles from the town. The old wells still had a life of their own, which imperiled the city.

The life of data is similarly long. Earth sciences data is used over and over, through exploration, appraisal, development, production and secondary/tertiary recovery.

The example of a drilled well is taken. It is a critical asset of the oil company. It may be used for several decades and may present a hazard for many decades more as demonstrated in the Kansas explosion incidents.

- The well needs to satisfy a large number of requirements: its trajectory needs to be correct. If the well has doglegs, running the casing will be a problem.
- If the well is not positioned correctly in geological and reservoir engineering terms, future production will not be optimized.
- If the well crosses a lease line, the oil company will not be allowed to produce the well.
- If the borehole condition has drastically deteriorated, production will be greatly affected and accurate description of the rocks may become problematic.

Still, little is said about the requirements on data, while the prime objective of the oil company is to drill the well as fast as possible.

6.6 SUMMARY

- Data can be biased by human intervention.
- Data vendors are not service companies.
- Data vendors deliver a product that will be used for many, many years.
- Data is a product, not a service.
- The goal of the data acquisition company is to deliver data that meets quality requirements, not the data users' expectations.

REFERENCES

[1] Allison, M. Lee, Bhattacharya, S., Buchanan, R., Byrnes, A., Nissen, S., Watney, L., Xia, J., Young, D., "Natural gas explosion in Hutchinson, Kansas: response to a geologic mystery," paper 2-1, Geological Society of America, North-Central Section – 37th Annual Meeting, Kansas City, Missouri, 2003.

[2] Bowker, G. C., *Science on the run*, MIT Press, Cambridge, 1994.

[3] Deffeyes, K. S., *Beyond oil, the view from Hubbert's peak*, Hill and Wang, New York, 2005.

[4] Horowitz, J. and Jurgens-Panak, M., *Total Customer Satisfaction-Lessons from 50 European Companies*, Pitman Publishing, London, 1992.

[5] Theys, P., "Le log," *The Log Analyst*, 1995.

[6] Theys, P., "Don't call them service companies," *Hart's E&P,* pp. 11-12, volume 79, n° 04, 4-2006.

[7] Theys, P., "Quality in the oil patch," *Petrophysics*, 2005.

7

Complexity

Tout ce qui est simple est faux, tout ce qui ne l'est pas est inutilisable.

Anything simple is false. Anything not simple cannot be used.

Paul Valéry (1867-1943)

The next two chapters deal with the increased complexity and complication that have affected the oil industry for the last thirty years. It is necessary to discern the difference between complexity and complication, and the related adjectives, complex and complicated (Table 7.1). Complexity implies systems and components that are unknowable. This means that they cannot be fully known. Complication and related complicated elements can be known with effort and time.

Table 7.1 Complex and complicated

Complex	Not simple, never fully knowable. Too many variables interact.
Complicated	Not simple, but ultimately knowable.

While nature is complex, the measurements to describe it are complicated. Complex situations are solved by complicated solutions. See Table 7.2.

Table 7.2 Complex situations and complicated solutions

Example	Situation	Simple situation	Complex situation
		Simple solution	**Complicated solution**
1	Mineralogy	One measurement	Multiple measurements
2	Positioning	Simple centralizers	Knuckle joints, etc.
3	Trajectory	Vertical	Deviated, horizontal
4	Invasion pattern	Three spacings	Many more spacings
5	Hole condition	Simple correction	Rugosity processing, etc.

The oilfield is looking for simple solutions, in the lines of the KISS [1] approach. Unfortunately, these solutions are not adapted to reality, which is often complex. Chapter 7 deals with increased complexity while Chapter 8 addresses complication.

7.1 HISTORICAL SUMMARY

7.1.1 From Colonel Drake to 1980

Before 1980, most drilled holes were vertical or had moderate deviation. The trajectory had a large radius of curvature. Many fields, especially in the Middle East had quasi-flat formation beds. A small variety of mud types, mainly based on water, were used: bentonite and barite were the preferred components. The mud invasion profile was cylindrical with the axis in coincidence with the one of the borehole. Reservoir targets were often thick beds.

7.1.2 From the 1980s

The advent of highly deviated boreholes, even horizontal, with smaller radius of curvature starts in the 1980s. New kinds of reservoirs arc investigated. A large variety of muds are introduced. It is obvious that the increased complexity is not the result of some form of masochism. Non-vertical wells traverse more pay zones or investigate more territory in exploration.

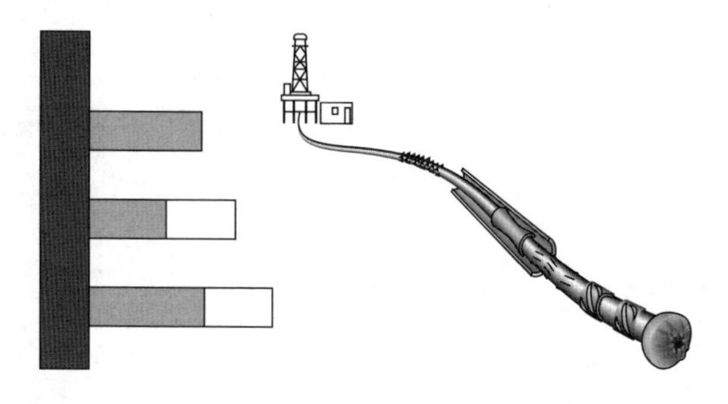

Figure 7.1

The increasing complexity of well data acquisition. On the left, a straight hole traverses thick horizontal layers. On the right, a logging-while-drilling string penetrates the rocks at an angle.

1. Keep it short and simple.

7.2 BOREHOLE TRAJECTORY AND SHAPE

7.2.1 Horizontal wells

The 1980s have seen the irresistible ascension of horizontal drilling (Table 7.3). Though they have been proven to increase production, horizontal wells present some specific challenges to formation evaluation data collection. Invasion patterns, tool positioning, hole size and shape are not understood in horizontal wells as completely as in vertical wells.

Table 7.3 Change from vertical to horizontal: number of horizontal wells

Year	1985	1986	1987	1988	1989	1990	1991	1992	1993	1994	1995
Number	6	40	63	145	270	1,063	1,325	1,475	1,625	1,765	1,950

By courtesy of Sperry Sun.

7.2.2 3-D trajectories

In addition to increased deviation, well trajectories are no longer smooth and regular. Thanks to geosteering, drilling keeps the well in the reservoirs (Fig. 7.2). Such shapes make the movement of the logging tools irregular. Also, the application of the sensors against the formation is not guaranteed.

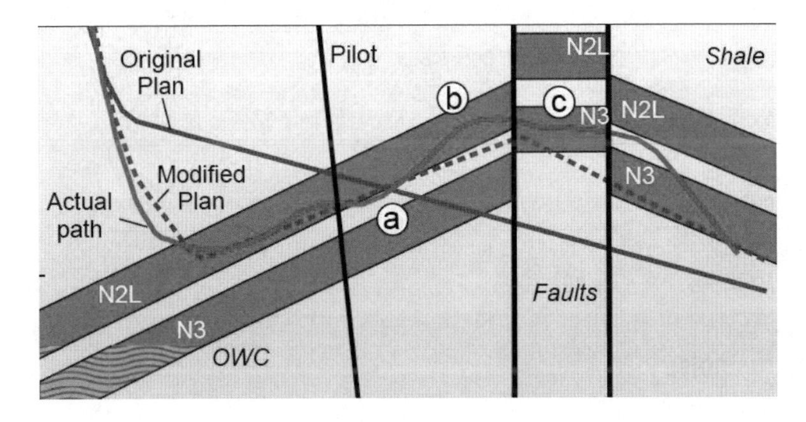

Figure 7.2

Well trajectories correspond to high curvature and numerous doglegs.

7.2.3 Multilateral wells

Early wells had a single stem. With multilateral wells, several branches start from a common trunk (Fig. 7.3). Measurements in any of the branches are problematic. Because of the increased productivity, this type of wells may become quite common in the future.

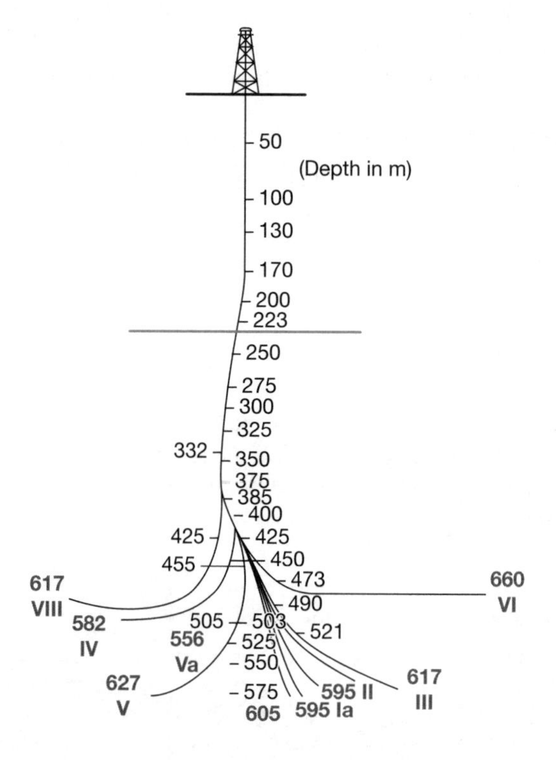

Figure 7.3

Multilateral well. As many as nine branches originate from the trunk well. By courtesy of Schlumberger.

7.2.4 Large holes

As a result of cost-cutting, pilot holes with small bit sizes have been often suppressed from the drilling programs. Large holes may be drilled from surface down to deep locations. Wireline open-hole logging tools external diameters vary from 3 to 4 in. With a 1.5-in standoff, the tool is not constrained to stay against the formation in vertical wells. Therefore, the position of the tool in the hole is not unique and cannot be known. Consequently, the borehole correction cannot be performed accurately. On Fig. 7.4, the tool is in the center of the hole (position 1), against the formation (position 2), or in-between (position 3). In position 1, the borehole correction will be large; in position 2, the smallest; in position 3, intermediate.

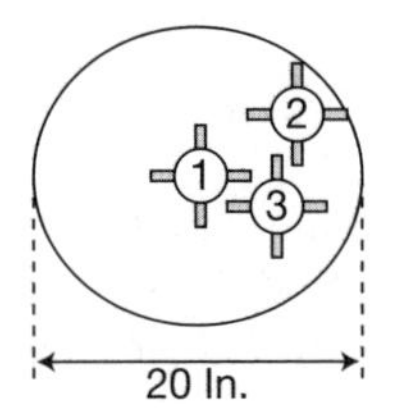

Figure 7.4

Positioning of a stood-off tool (1.5-in standoff) in a large hole cannot be completely known. Positions 1, 2 or 3 are possible.

7.2.5 Invasion patterns

Going from vertical wells to a different trajectory has a direct impact on the invasion pattern. On Figure 7.5 (left), a vertical hole is seen cut by a horizontal plane. The tool is well centered in the borehole. Little space is occupied by the mud, which minimizes borehole effects. The invasion zone is cylindrical. Its axis coincides with the axis of symmetry of the logging tool. Environmental corrections are small and manageable. On Fig. 7.5 (right), a horizontal well is observed as cut by a vertical plane. The tool, part of the bottom hole assembly, belongs to a string that wobbles and makes a spiral hole (when cut by a horizontal slice, it shows as an ellipse). The invasion zone is affected by fluid segregation and does not display a simple cylindrical symmetry. Environmental corrections are difficult, if not impossible.

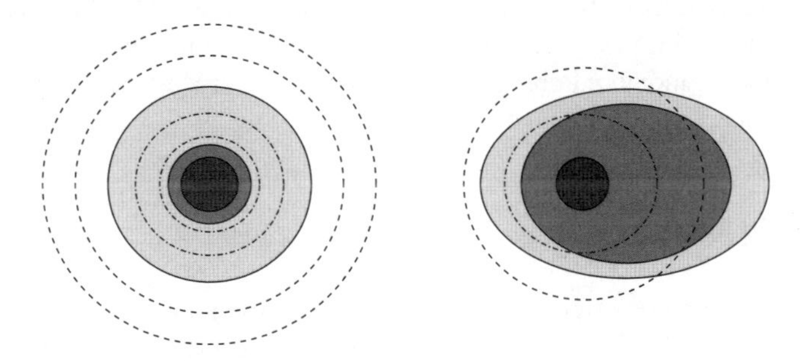

Figure 7.5

Invasion profile: left: a classical vertical well seen from above. Right: a horizontal well seen from the side.

7.2.6 Movement of logging tools and drilling assemblies

On non-vertical wells, accurate depth measurements are a challenge. Positioning a perforating gun or a formation tester with a wireline cable requires a correction for tool creep.[2] Drilling strings have even more complex movement.

2. See Chapter 15.

7.3 MUD COMPOSITION

Drilling muds have been modified to improve drilling and reduce caving and shale swelling. Unfortunately, the effect of these new muds on logging measurements could be detrimental. Two examples of muds affecting logs, formate cesium based and petrofree muds, are considered.

7.3.1 Formate Cesium

Cesium Potassium (CsK) formate muds are used to reduce formation damage. They are also more apt at being stabilized in terms of mud density. These muds are quite different from the known water- and oil- based muds (Table 7.4).

Table 7.4 Parameters for CsK formate mud

Parameter	CsK formate	Units
P_e	199	b/e
HI	0.545	
σ	148	c.u.

Laboratory studies and nuclear modeling are needed to quantify the effects of this mud to common logging measurements. The conclusion of this research [3] indicates that the measurements can be used with reduced accuracy and precision. In particular, significant invasion by CsK formate mud could obscure the presence of hydrocarbons through its effect on electrical and nuclear logs.

7.3.2 Petrofree Mud

Petrofree mud is a synthetic ester-based mud. When it is used, unexpected large separations on the density-neutron plot can be observed, much like a gas effect in zones with no gas [2]. Recent lab measurements, under reservoir pressure and temperature conditions, have revealed that petrofree mud can absorb gas. It raises an array of invasion scenarios that explain the log response in liquid-filled reservoirs and gas reservoirs. Modeling results reveal that the effects of mud invasion can be influenced by a number of factors, including absorption of gas by the mud. The puzzling aspect of petrofree mud is that its strange effects are not systematically observed.

7.3.3 Other muds

These two examples are not unique. Many different mud types and compositions have been introduced in the oilfield in the recent years. Most of the time, the effects of muds on measurements have not been characterized, unlike the two previous examples that have attracted considerable attention. The bigger picture consists in assessing the gains in drilling efficiency and the losses in accuracy and precision of the measurements in order to clarify the economical impact of these new muds.

7.4 COMPLEXITY INDUCED BY COMBINED MEASUREMENTS

If the movement of a relatively short tool in a vertical and regular hole can be understood, this is not the case for a long tool moving in a deviated hole of irregular size. The tool may bend. The positioning of the tool is difficult to model and predict. Incompatible positioning schemes for different measurements make it even more difficult. It is possible to have in combination a tool that needs to be centralized and another one that requires to be applied against the formation.

Long wireline tools with sensors distant one from another, may investigate different formations if the rocks present some azimuthal variations (Fig. 7.6). The issue has been investigated in the context of consonance [1]. An interpretation combining sensor A seeing the "white" zone, sensor B looking at the grey zone and sensor C, in front of the "black" zone would give odd results.

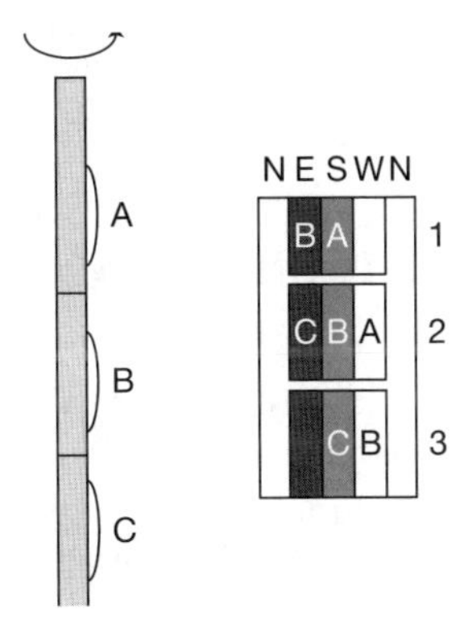

Figure 7.6

Rotating tools in a formation with poor azimuthal continuity. The formation is made of three rock types (black, grey and white, 120° apart). The three tools A, B and C are aligned. As the tool string rotates, the three tools see different formations. Over interval 2, for instance, tool A sees the white rock, the memorized tool B sees the grey rock, the memorized C tool sees the black rock. Several logging passes may be needed to better understand the formation.

7.5 COMPLEXITY INDUCED BY THE HOLE SHAPE IN THE LOGGING-WHILE-DRILLING MODE

While it is possible to visualize the impact of hole size and shape at a given depth on the measurements performed at the same depth, it is less understood that stabilizers a few feet away from the detector (or receiver) may also strongly affect the measurement by creating a standoff (Fig. 7.7).

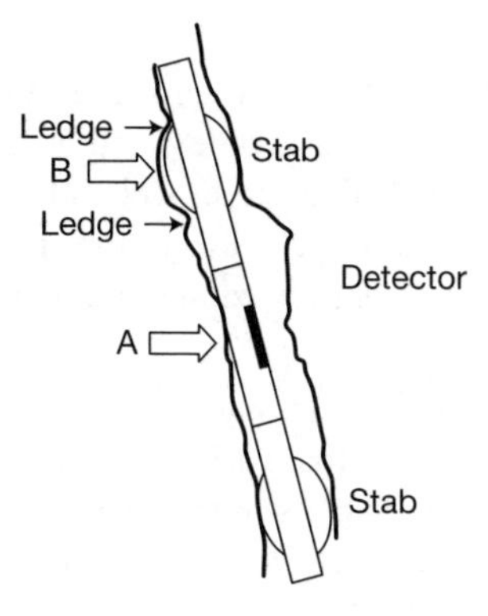

Figure 7.7

The LWD detector is not only affected by the borehole shape in front of the detector (A), but by the hole shape in front of the stabilizers (B), denoted by "stab".

7.6 INCIDENCE OF COMPLEXITY ON DATA ACQUISITION

As compared to the data required to describe a vertical well, the quantity of information needed in a modern well is multiplied several times. More complexity implies an increased need for information. The positioning of the logging tools can be defined with accelerometers and the measurement of orientation (relative bearing, azimuth, deviation). Exotic muds need to be better documented so that the effects are understood.

Because of the complexity, the logging companies need to communicate much more information to the data user.

7.7 SUMMARY

- Nature is complex.
- Wells have become more complex with the introduction of highly deviated and horizontal wells.
- The solutions to complex situations are complicated.
- Complexity needs to be managed. It should not be ignored or brutally simplified.

REFERENCES

[1] Andreani, M., Klopf, W., Casu, P. A., "Using consonant measurement sensors for a more accurate log interpretation," SPWLA 39th annual logging symposium, 1998.

[2] Badruzzaman, A., Sheffield, A.J., Adeyemo, A. O., Logan, Jr. J. P., Stonard, S. W., "The ubiquitous neutron/density tool response in petro-free mud: New insights to addressing unresolved issues," SPWLA 46th annual logging symposium, 2005.

[3] Pedersen, B.K., Pedersen E.S., Morriss, S., Constable, M.V., Vissapragada, B., Sibbit, A., Stoller, C., Almaguer, J., Evans M., Shray F., Grau, J., Fordham, E., Minh C.C., Scott, H., McKeon D., "Understanding the effects of Cesium/Potassium formate fluid on well log response-a case study of the Kristin and Kvitebjorn Fields, offshore Norway," paper 103067, SPE annual technical conference and exhibition, San Antonio, 2006.

8

Complication

Complexity, uncontrollable, is difficult to grasp, but complication is built and designed by the data vendors. Complication is sometimes necessary. In all cases, its digestion needs to be facilitated to oil companies and to data users.

8.1 INCREASED COMPLICATION

Complication is now directly felt by the data user in the following ways:

- Delivery is a very intricate process.
- There is now a large number of tools. They are not completely compatible. They are not delivering the exact same information.
- These tools are quite sophisticated. The underlying physics are beyond the training of many geoscientists. Sonic dispersion diagrams, pressure derivatives curves and optimized spin polarization schemes are a few examples of topics that require expert attention.
- There is a parallel development of response and environmental corrections for each of these tools. The documentation is overwhelming, though not complete. This information cannot be managed on paper any longer (and will soon be only available in digital format).
- Modern tools have capabilities increased thousand times, as compared to their ancestors.
- Data deliverables have huge volume, not adapted to easy database management. Information on the digital deliverables is limited.
- With the introduction of logging-while-drilling, time base data makes its entrance on the stage. How can it be managed?

8.2 COMPLICATION IN DELIVERY [1]

8.2.1 Simple beginnings

A historical recap of the measurement evolution is useful. The first tool was made of four electrodes tied up on a bakelite shaft. The first log was displaying one parameter in function of depth. The measurement involved a single auxiliary parameter, the constant K_l, at that time equal to 7.52 (Fig. 8.1).

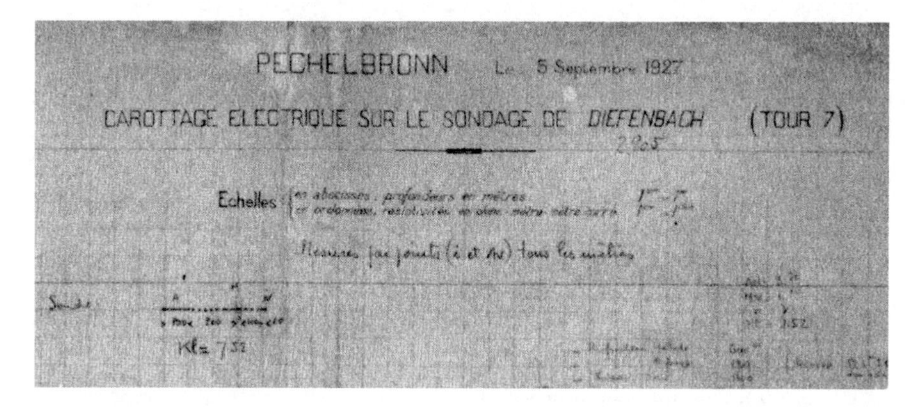

Figure 8.1

The sonde characteristics (K_l = 7.52) are shown on the first log.
By courtesy of Schlumberger.

The first log was delivered as a piece of paper. It contained all the information available to the crew. In particular, Henri Doll, the logging engineer, added his name to the print, starting a long tradition that recognizes the creator of the log. There was no separate delivery for the data producer and the oil company (Compagnie pétrolière de Péchelbronn). There was no digital data that could have been stored under a different support.

8.2.2 Transition to digital data

This delivery process did not change much till the late 1970s: one document was delivered at the end of the logging job – a film (and possibly a few prints directly copied from the film). The delivery was performed before the logging tool was rigged down, minutes after the end of the periods of measurement. There was no opportunity for playback. If the measurements were not satisfactory or the film poorly developed, it was necessary to repeat the

1. This section comments on the current delivery. Chapter 14, entitled "deliverables" is expanded on recommended delivery in the future.

run. But, again, the resulting deliverable would be presented immediately to the oil company representative. There was no waiting, no telephone call and no email to insist on final delivery. There was generally no draft and no final copy. The film delivered at the well site was the unique and final deliverable.

In the mid-1970s, part of the information was recorded on tapes, but the film/print was still the main document. A binary tape recorder was put next to the analog camera equipped with galvanometers that drew curves on a film. The tape recorder was handling five channels and was not capturing any auxiliary data. The digital tape could not be used unless the film was available and contextual data extracted from it.

Computers appeared in the logging unit in the late 70s and started encrypting non-sampled data onto a tape. Tapes became the principal component of the deliverables. Several tape formats, and the resulting complication appeared on the market.

8.2.3 21st century deliverables

Today, logging companies deliver data encapsulated in several subcomponents:

- A graphical file, reminiscent of the print,
- Several digital files submitted to the oil companies as compact disks or DVDs. The data can also be transmitted directly to a digital database without any transition on a physical support.

The digital data set does not completely overlap the graphical data set. Some critical information is available in the graphical file, but not in the digital files. The converse is also true. The data user still needs at least two records, including the graphical file.

8.3 MULTIPLICITY OF FORMATS

8.3.1 Setting the problem

Digital records are structured by digital formats. Digital formats are not standardized. Ironically, it may be stated that the nice thing in the industry today is that there are so many standards to choose from.

Because of the lack of standardization, it is not rare to misname data objects. Errors of content, unit and origin are common. In addition, a lot of data users work on a very limited set of depth- or time- sampled data gathered in a composite file of multiple runs with very little contextual information. Calibration data, depth system information, signal processing options are seldom captured in the project database. A number of popular formats are reviewed in the next sections.

8.3.2 RP 66/DLIS

RP 66 is the acronym for Recommended Practice 66, a document built for the American Petroleum Industry, API [3]. There are several implementations of RP66 (Geoframe Archive format, GeoShare, POSC Exchange Format – based on the EPICENTRE data model, RODE – BP-developed Record Oriented Data Exchange for encapsulating original geophysical SEG-A/B/Y [2] files).

DLIS stands for Digital Log Interchange Standard. Specifications for this format can be found in the RP66 document. It was introduced as an enhanced exchange format for well data. The primary features of RP66 V1 are machine independence, self-description, semantic extensibility and efficient handling of bulk data.

If the format is well designed, it does not impose any specific content. The content is often assumed to be complete and relevant by the data user. A number of convenient features, such as the use of attributes for the naming of data objects, have not been fully implemented by data vendors.

8.3.3 LAS

LAS, was introduced by CWLS, the Canadian Well Logging Society with the following vision statement [2].

The CWLS's Floppy Disk Committee has designed a standard format for log data on floppy disks. It is known as the LAS format (Log ASCII Standard). The LAS record consists of files written in ASCII containing minimal header information and is intended for optically presented log curves... The purpose of the LAS format is to supply basic digital log data to users of personal computers in a format that is quick and easy to use.

LAS version 3.0 encompasses all domains of information, but most practical field records are delivered in the 2.1 version (as of the printing of this book). The LAS format is popular because of its high usability. The creators admit that it contains "minimum information." This statement is often forgotten and can offer some risk when complicated measurements are acquired.

8.3.4 Other formats

- TIF, a method to stream a record-oriented, is a data exchange for Unix or PC systems. It is not a self-navigating format.
- LIS is phased out because it is not Y2K compliant.

All formats are for data exchange. They are used to flow data from the acquisition site to the oil company database. Each oil company has a data model that selects data objects from the datasets so provided. It is rare that all data provided by the data acquisition company is stored in the corporate database.

2. Almost each major oil company has at least one SEGY version.

8.3.5 Graphical formats

- PDS is a Schlumberger specific graphical format that mimics the print file of the early logging days. PDSView is a freeware that enables an easy conversion to GIF, CGM and TIFF.
- TIFF is an abbreviation of Tagged Image File Format. Much use by Western Atlas, it can be read by a number of freeware.

In addition, there is now a plethora of graphical formats, corresponding to the development of digital still and movie cameras. It is likely that logging companies will use these formats (e.g., CGM and PDF, well described in Wikipedia), in the design of future deliverables.

8.3.6 Other digital platforms

Data can be transmitted from the well site to a decision center, especially for real time decisions. Data is no longer frozen or packaged in an archive, but it can flow continuously. The associated platforms are useful, but present some risks, if data is updated or corrections modified. A strict recognition of the different versions of the data needs to be kept.

- WITSML, the Well site Information Transfer Standard Markup Language is used for transmitting technical data between organizations in the petroleum industry.
- Openspirit is a set of middleware [3] applications that promotes full integration in data management. [4] Openspirit applications definitely improve the transfer recognition and use of data, **when the data does exist**. It becomes ineffective when the original content has been incomplete or inappropriate.

8.4 CONTENT OF THE DIGITAL RECORDS

Present well deliverables take multiple forms and include a graphical print and a digital record. The digital record is produced in a variety of formats, as previously discussed. The data vendor may store (for a short period of time) the related "proprietary" data and deliver the "customer" data. Figure 8.2 illustrates the complication.

3. Middleware is a form of software that connects applications. The software consists of a set of services that allows multiple processes running on one or more machines to interact.

4. The development of the OpenSpirit Application Integration Framework began in 1997, when a consortium of oil companies and software vendors sought to address the inefficiencies and high costs related to the poor integration capabilities of Exploration and Production applications and their related data stores.

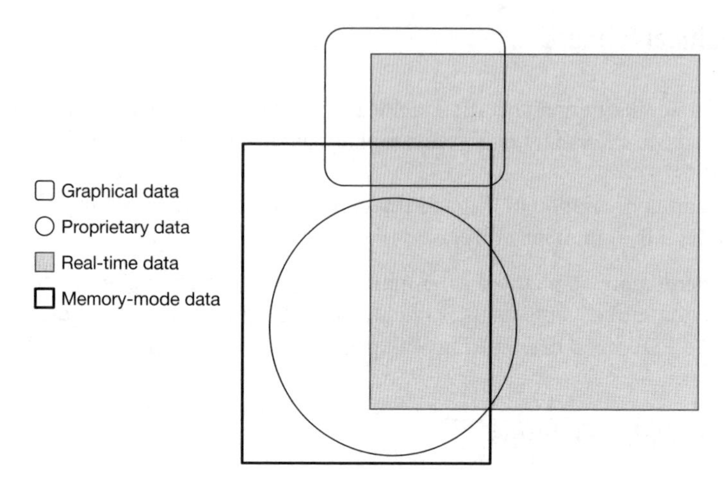

Figure 8.2

A simple attempt to describe logging job deliverables. In this example the conveyance is on drill pipes. There is no strict coincidence between the different types of data.

8.4.1 Classification of data objects

The digital records list parameters, sampled channels (vectors) and sampled arrays. The files vary extensively in size. Some logging companies have introduced a classification in order to distinguish the proprietary nature of the data objects (Table 8.1).

Table 8.1 Classification of digital data objects (Schlumberger)

Type	Purpose
Basic	Objects that are shown on a graphical file.
Customer	All objects required to reconstruct the main curves.
Producer	Life of the tool.
Internal	Erased as soon as the surface system is shut off.

This classification is not rigorous as some customer objects are of no interest to the data user, while some producer data is critical for the reconstruction of log (if a calibration is incorrect, or if the environmental corrections have erroneous inputs).

Table 8.2 gives an example of a subset of the parameters required for a cased-hole monitoring job. It is likely that only the tool designer can understand each specific parameter. The test could be done on CDDE, the RSC last command data echo. Interestingly enough, some quantified uncertainties (e.g., BSAL_SIG in Table 8.2) are listed as "customer" deliverables.

Table 8.2 Parameters describing a reservoir monitoring measurement

Type	Mnemonics	Description
BASIC	BSAL	Borehole Salinity
CUSTOMER	ACOR_SIG	RST Far Average Carbon/Oxygen Ratio
CUSTOMER	AIRB_DIAG	Uncertainty
CUSTOMER	AQTF	RST Air Borehole Diagnostic
CUSTOMER	AQTM	RST Far Frame Acquisition Time
CUSTOMER	BADL	RST Near Frame Acquisition Time
CUSTOMER	BADL_DIAG	RST Bad Level Flag, Sigma Phase 1
CUSTOMER	BSAL_SIG	RST Bad Level Diagnostic
CUSTOMER	BSFL_DIAG	RST Borehole Salinity Uncertainty
PRODUCER	BECD	RST BSAL Filter Level Diagnostic
PRODUCER	BESE	RST Beam Current Set Point
PRODUCER	CACU	RST Beam Current Setting
PRODUCER	CAV	RST Cathode Current
PRODUCER	CDCE	RST Cathode Voltage
PRODUCER	CDDE	RSC Last Command Data Echo

8.4.2 First example: the gamma ray log

The gamma ray curve, GR, is one of the simplest and most common. It can be used to qualitatively analyze the volume of shale. It has also an essential role in the depth-matching process: When run with every logging tool, it enables the different logging curves to be put in depth with a reference curve.

In 1999, the number of traces describing a GR by Schlumberger was already quite high:

gr, gr1, gr2, gr2_sl, gr3, gr4, grbc, grc, grdn, grfc, grin, grlt, trm1, grn, grp, grr, grra, grrc, grrt, grsg, grt, grt1, grt2, grt3, grt3, grtc, grte, grup, grow, gr_cal, gr_cal, gr_dn_raw_d, gr_lt_raw_d, gr_rt_raw_d, gr_sl, gr_up_raw_d.

A decade later, the number has been multiplied many times over. The data user likes to understand why so many different names exist and what they mean.

8.4.3 Second example: a modern combination of measurements

Figure 8.3 shows an example of digital deliverable for a combination of resistivity (ARC5 tool, measuring resistivity) and density (ADN or azimuthal density neutron tool). The density-neutron counts 13 1.2-in. sampled channels and 98 6-in. sampled channels [5]. The user definitely needs some help to fully utilize the data set. Quite often, the definition of the mnemonics used in the digital deliverable is not readily available.

5. The numbers of channels in this example are conservative. It is possible to observe more than 100 1.2-in sampled channels and more than 200 6-in sampled channels for ADN only.

Channels		File: MWD_10.039			Sequence: 32	
Origin: 43						
ARC5-825: 8.25-in. Array Resistivity Compensated						
Spacing: 1.2 in				**Number of Channels: 2**		
TICK_ARC_GR		TICK_ARC_RES				
Spacing: 6.0 in				**Number of Channels: 108**		
A112	A114	A122	A124	A132	A134	A142
A144	A152	A154	A16B	A16B_UNC	A16H_UNC	A16L_UNC
A212	A214	A222	A224	A22B	A22B_UNC	A22H_UNC
A22L_UNC	A232	A234	A242	A244	A252	A254
A28B	A28B_UNC	A28H_UNC	A28L_UNC	A34B	A34B_UNC	A34H_UNC
A34L_UNC	A40B	A40B_UNC	A40H_UNC	A40L_UNC	AGRACQTM	AGTM
APRS	AR12	AR14	AR22	AR24	ARESACQTM	
ATAT_ARC_IMG	ATMP	BATV_ARC	ECD_ARC	GR_ARC	GR_ARC_CAL	
GR_ARC_FILT	GR_ARC_RAW	ISBD	P112	P114	P122	
P124	P132	P134	P142	P144	P152	P154
P16B	P16B_UNC	P16H_UNC	P16L_UNC	P212	P214	P222
P224	P22B	P22B_UNC	P22H_UNC	P22L_UNC	P232	P234
P242	P224	P252	P254	P28B	P28B_UNC	P28H_UNC
P28L_UNC	P34B	P34B_UNC	P34H_UNC	P34L_UNC	P40B	P40B_UNC
P40H_UNC	P40L_UNC	PFPG_ARC	PPPG_ARC	PR12	PR14	PR22
PR24	SHK1_ARC	STAT_ARC	STAT_ARC_SUM	TAB_ARC_RES	TEMP_ARC	
ADN8-AA: 8.25-in. Azimuthal Density Neutron						
Spacing: 1.2 in.				**Number of Channels: 13**		
DRSI	DSAM	LSAZ_P1F	NSAM	PESI	RLSI	ROSI
RSSI	SAZ1_P1F	SAZ2_P1F	TICK_DEN	TICK_NEU	USI	
Spacing: 6.0 in.		**Number of Channels: 98**				
DCAL	DPHB	DPHL	DPHR	DPHU	DRHB	DRHL
DRHO	DRHR	DRHU	DRRT	DRSC	EFRA	FAR1
FAR2	FAR3	FAR4	FAR5	FAZ1_R	FAZ2_R	FAZ3_r
FAZ4_R	FAZ5_R	HEHV	HORD	IBT1_ADN	IBT2_ADN	LSAZ
LSHV	LSW3	LTBV_ADN	NAZ1_R	NAZ2_R	NAZ3_R	NER1
NER2	NER3	P20V_ADN	PEB	PEF	PEL	PER
PESC	PEU	RHOB	RHOL	RHOS	RLSC	ROBB
ROBL	ROBR	ROBU	ROLB	ROLL	ROLR	ROLU
ROSB	ROSC	ROSL	ROSR	ROSU	RPM_ADN	RSSC
SAZ1	SAZ2	SHK1_ADN	SHK2_ADN	SOAB	SOAL	SOAR
SOAU	SOIM	SONB	SONL	SONR	SONU	SOXB
SOXL	SOXR	SOXU	SSHV	SSW1	SSW3	TAB_DEN
TAB_NEU	TAZN	TNPH	TNPH_UNC	TNRA	TNRA_UNC	TTEM_ADN
U	UB	UL	UR	USC	UU	VERD
System and Miscellaneous						
Spacing: 1.2 in		**Number of Channels: 2**				
TDEP; 1		TIME; 1				
Spacing: 6.0 in.		**Number of Channels: 15**				
BS	DPHI	FF	FREQ	HDAV	PHIT_B	R0
ROP1_RM	ROP5_RM	ROP_RM	RWA_B	TDEP; 2	TEMP	TIME; 2
TVDE						

Figure 8.3

List of channels for the ARC5-ADN combination.

8.4.4 Composite logs

Because of more intricate trajectories, from vertical to high-deviation and horizontal shapes, a well often requires several bit runs before completion. This means that several measuring tools and several types of mud are going to be used in sequence. For each run, there is a separate set of parameters. For instance, different tools call for different calibration parameters and different muds require various environmental correction processing, one per run. As an example, the Schlumberger data set for a well contains one separate DLIS file per run, one WSD (Well Site Data), one set of parameters and one complete set of channels (customer/basic) per run. A so-called composite data record (Fig. 8.4) cannot contain individual run information.

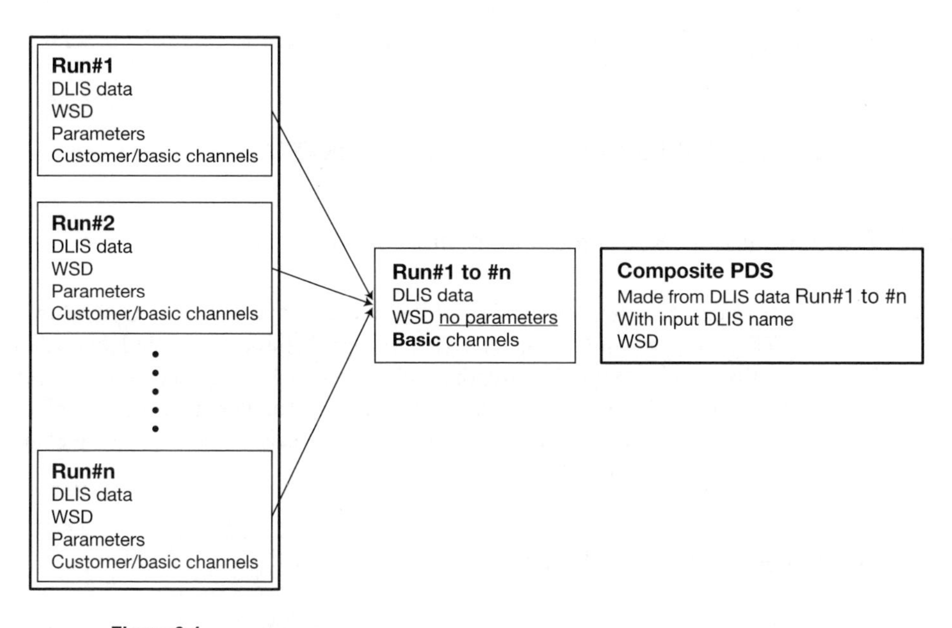

Figure 8.4

Example of multiple runs.

The composite DLIS record in the center of the figure does not contain any parameter describing individual runs. In addition, auxiliary curves, so useful to validate the data, are absent.

The following strategy is recommended. Individual runs need to be used for quantitative petrophysical interpretation. A composite run may be delivered for correlation work.

8.4.5 Variability in the volume of delivery

Another difficulty encountered by the data user is the lack of consistency in the number of delivered channels. Table 8.3 lists the number of channels delivered for logging-while-drilling density jobs in 36 different wells in the same field. This number varies from 2 to 165. There are 17 different numbers of channels. Intuitively, it is guessed that the low number corresponds to a simple, low cost product, with two basic curves. Conversely, the high number should be linked to a record containing images. It is still hard to explain why there are 15 additional intermediate values. The multiplicity of values does not help spot and identify missing information.

Table 8.3 Variability in delivery. In this field study, 36 wells have been surveyed with a density log. In parallel, a digital record including channels has been delivered for each well. The number of channels varies between 2 and 165.

Well	1	2	3	4	5	6	7	8	9	10	11	12	13	14	15	16	17	18
Channels	2	2	2	2	72	72	68	68	2	2	89	89	2	2	89	114	114	152

Well	19	20	21	22	23	24	25	26	27	28	29	30	31	32	33	34	35	36
Channels	152	152	152	111	111	149	149	149	151	114	73	76	118	62	165	74	49	154

8.5 ADDITIONAL ISSUES RELATED TO DATA CONTENT

8.5.1 Further division: Real-time and memory mode

Since the early development of logging-while-drilling, it has been possible to send information in real time. The rate of transmission, supported by signals through the mud, considerably limits the quantity of data. After drilling, the tools come back to surface and a detailed data set is transferred from the memory of the tool. For the same depth, two data objects, $x_{real\text{-}time}$ and x_{memory}, may coexist for the same parameter x. Because the processing chains are not identical, the two values are likely to be different.

8.5.2 Several versions of a job may exist

Today, data products delivered at the well site are not final. A "draft" copy, necessary for immediate decisions is generally proposed to the oil company. It may happen that the field engineer recognizes a mistake and produces an "intermediate" copy. A few days, or a few weeks later, a final copy is delivered. The multiple versions may contain slightly different information.

An example of incident created by poor housekeeping of the multiple versions is given:

A field engineer performs an open-hole job. He delivers a "draft" version which is instantly transmitted to a team of geoscientists, located offsite. They work intensively to interpret the data and provide perforating intervals. Meanwhile, the field engineer discovers that the stretch correction has not been applied to depth. He plays back the data, this time with a correct depth, 10 ft off-depth from the previous data set. This field engineer is replaced by a perforating engineer and the interpreted data and the perforating intervals are transmitted to the rig. The perforating engineer uses the corrected data set, available at the well site, to position the perforating guns. As a consequence, the perforations are off-depth by 10 ft.

8.5.3 Changes of formation parameters with time

In most cases, a single set of formation characteristics is delivered for a given depth. In a logging-while-drilling situation, the same depth interval is observed by the measurement equipment several times. Assuming four drilling runs, the shallow formations depicted on

Fig. 8.5 are surveyed eight times, four times down and four times up. In real conditions, a BHA, or bottom hole assembly, loaded with measuring sensors, passes in front of a rock formation many more times. Opening and closing fractures, linked to change of mud weight and borehole pressures can be monitored.

As a result, not one, but several values can be attributed to a depth index for a given log. Time-indexed files then become valuable. They must be stored accordingly, in databases that generally assign one parameter per depth index.

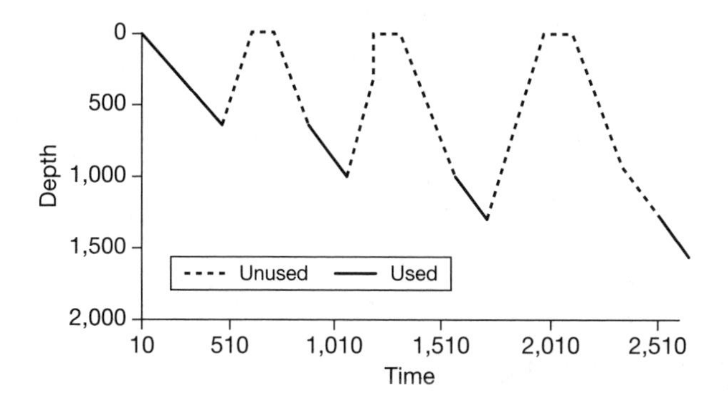

Figure 8.5

The same formation is observed several times by the LWD tool.

Logging-while-drilling has allowed some insight in the behavior of formation parameters. The wireline logging technique gives the feeling that rock parameters do not change with time. This is not surprising as wireline passes are performed within a short time span. These passes take place after mud has been circulated and the well has been stabilized. Before drilling, the formation was indeed in a stable situation. Drilling and mud invasion submit the rocks to a traumatic experience. Wireline logging takes place in more peaceful conditions. Nevertheless they are not likely to be the conditions existing before or long after drilling as schematized in Fig. 8.6.

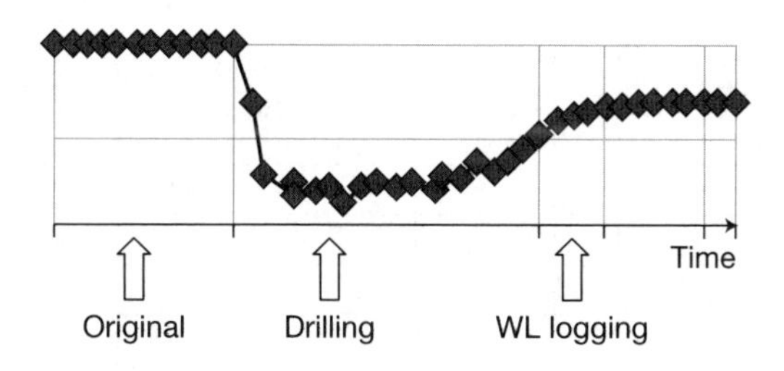

Figure 8.6

Evolution of a formation parameter with time.

8.5.4 The seven dimensions of a data object

The complication of logging data can be visualized in Table 8.4. A simple data object can be represented with at least seven dimensions.

Table 8.4 Dimensions of data

Attribute	Status 1	Status 2	Status 3	Status 4
Confidentiality	Proprietary	Customer		
Processing	Raw	Calibrated	Corrected	Filtered
Pass	Main	Repeat		
Acquisition	Real time	Memory		
Dimension	Time base	Depth base		
Delivery	Draft	Final	Intermediate	
Presentation	Graphical	Digital		

8.6 SOPHISTICATION IN MEASURING TOOLS

The first logging tools wcrc madc of a mandrel with few coils of wire added to it. Today, they are transformed into mountains of technology. At the beginning, they could still look like a piece of pipe, but now, they are housing high-technology electronics. Slowly emerging out are detectors, acoustic hydrophones and caliper arms. The acoustic tool, measuring sonic transit time is an example [1]. At the beginning, it had a couple of transmitters. It then evolved in a costly device with eight hydrophones (Table 8.5). The SonicScanner, in its most common and popular form records 90 times more data than the first sonic digital tool.

Table 8.5 Increasing complication of the Schlumberger sonic tool
By courtesy of A. Brie.

Type of measurement	Modes	Stations	Hydro-phones/ station	Wave-forms/ station	Wave-forms	Samples/ waveform	Total samples
Analog Borehole Compensated Sonic	1	4	1				
Long Spaced Sonic Digitized	1	4	1	1	4	512	2,048
Array sonic	1	8	1	1	8	512	4,096
Array Sonic – Multi Gain	1	8	1	2	16	512	8,192
Dipole Sonic Imager							
DSI – Lower dipole, Upper dipole, P&S, Stoneley	4	8	4	1	32	512	16,384
DSI – BCR, P&S, Stoneley	4	8	4	1	48	512	16,384
Sonic Scanner							
Sonic Scanner Concise	6	13	8	8	182	512	76,544
Sonic Scanner Record All Data	6	13	8	8	416	512	186,368

8.7 SUMMARY

- The complexity of nature is followed by an explosion of complicated data products.
- There is a multiplicity of formats and a huge variability in content and volume of delivery.
- Data content is poorly described and documented.
- Complicated data sets are difficult to quality control, and hence difficult to fully use.
- The lack of standardization makes the database management even harder.

REFERENCES

[1] Brie, A., personal communication.

[2] *LAS version 2.0, a floppy disk standard for log data*, the Canadian Well Logging Society, Calgary, 1992.

[3] *Recommended Practices for Exploration and Production Data Digital Interchange: API RP 66, V2: Second Edition,* American Petroleum Institute, 1996.

9

WYSINWYTII

Some complication may be needed and impossible to avoid. But the oil industry has been faced for years with logging deliverables that are sometimes misleading. The ambiguities have led the users to make erroneous assumptions on what the data means. Some examples are reviewed.

The chapter has a barbaric title which means: What You See Is Not What You Think It Is, by opposition to the user-friendly screens proposed by well designed software. The delivered log components may not exactly represent what it appears.

9.1 EARLY PERMEABILITY CURVE

Ambiguity started early with the presentation of permeability on paper (Fig. 9.1). The user could be confused and believe that permeability was actually measured. The scale unveils the partial deception. It is expressed in millivolts (mV) and not in millidarcies (mD). The curve shown was, in reality, a spontaneous potential curve.

9.2 LOG HEADERS

Modern log headers are confusing (Fig. 9.2). Bit size is unambiguous. The information comes from the driller and is verified at surface. Caliper is well named: a measurement of one axis of the borehole. An erroneous curve title would have been "borehole diameter." Density correction is acceptable and does not try to confuse the data user.

"Formation density" and "formation Pe" are definitely misleading. The "formation density" curve results from a complicated processing, starting with the detection of gamma ray counts, the use of spectroscopy to identify the energy of the incident gamma rays, a sophisticated processing to remove mud cake and hole effects (processing based on a model that necessarily simplifies the real geometrical relationship between the borehole and the formation), a transform between electronic density and mineral density, etc. The curve does not represent "formation density." The same comment applies to "formation Pe."

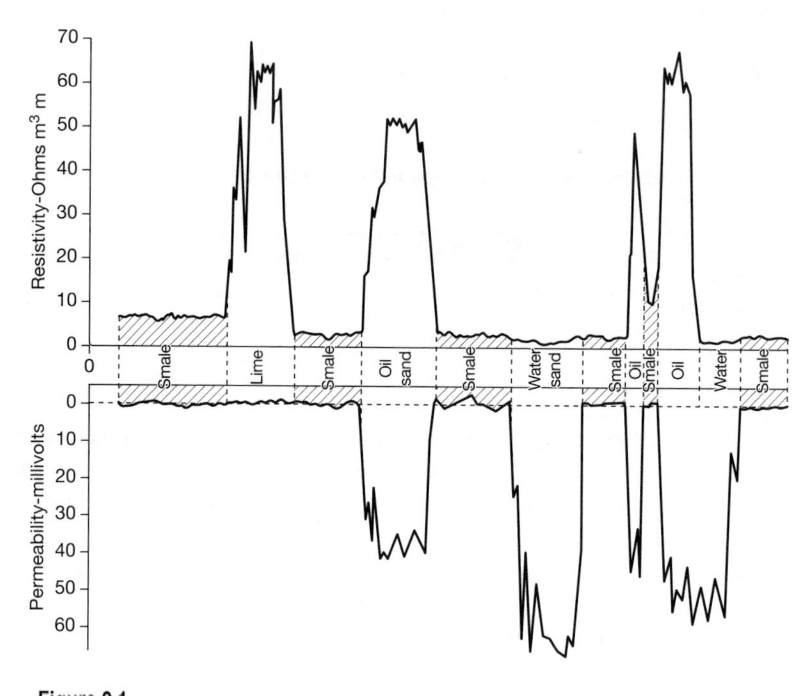

Figure 9.1

Presentation of an early log.
By courtesy of MIT press.

The argument is not pedantic. The data user may ask: "What is the difference between measured density and formation density?" The difference is the uncertainty. Uncertainty is what needs to be properly managed. It should not be assumed to be null. In conclusion, the term "measured parameter" should always be preferred to "formation parameter."

<table>
<tr><td>HILT Caliper (HCAL)
10 (IN) 20</td><td></td><td colspan="2">Density porosity crossover</td></tr>
<tr><td>Bit Size (BS)
10 (IN) 20</td><td>Calibrated Downhole Force (CDF) (LBF)
– 200 1,800</td><td colspan="2">H. Res. formation density (RH08)
1.95 (G/C3) 2.95</td></tr>
<tr><td>Undergauge</td><td>H. Res. Density Standoff (DS08)
2.5 (IN) 0</td><td>H. Res. formation Pe (PEF8)
0 (...) 10</td><td>Density Correction (HDRA)
– 0.25 (G/C3) 0.25</td></tr>
</table>

Figure 9.2

Log insert explaining the different curves.

9.3 DEPTH

Figs 9.3 and 9.4 represent the information on depth extracted from two prints.

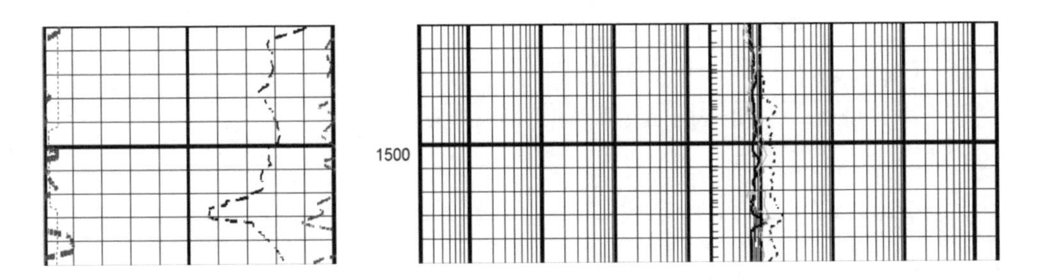

Figure 9.3

Depth number from a LWD recording.

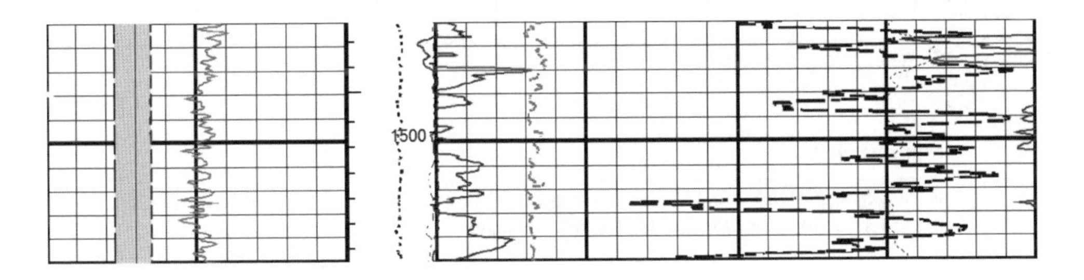

Figure 9.4

Depth number from a wireline recording.

The features seen in front of the depth number could be assumed to be at the same depth. They are not. As explained in Chapter 15, LWD depths are not stretch-corrected while the wireline depths are stretch-corrected. Does it matter? At 1,500 m, cable or drill pipe stretch can be evaluated at 6 to 7 ft (2 to 2.5 m).

9.4 VOLUME INTEGRATION

Logging calipers are used to perform a useful and simple task: compute the volume of the hole and anticipate the difference between the hole volume and the volume of the casing that will be drawn in the hole. The quantity of cement to be pumped can be inferred. Fig. 9.5 shows information on the meaning of the "pips," expressed in ft^3 or m^3. Fig. 9.6 shows examples of pips on the log. Fig. 9.7 shows the result of the volume computation.

PIP SUMMARY

⊦ Integrated hole volume minor pip every 10 F3
⊢ Integrated hole volume major pip every 100 F3
⊣ Integrated cement volume minor pip every 10 F3
⊣ Integrated cement volume major pip every 100 F3

Time mark every 60 S

Figure 9.5

Information on the pips used for hole volume integration.

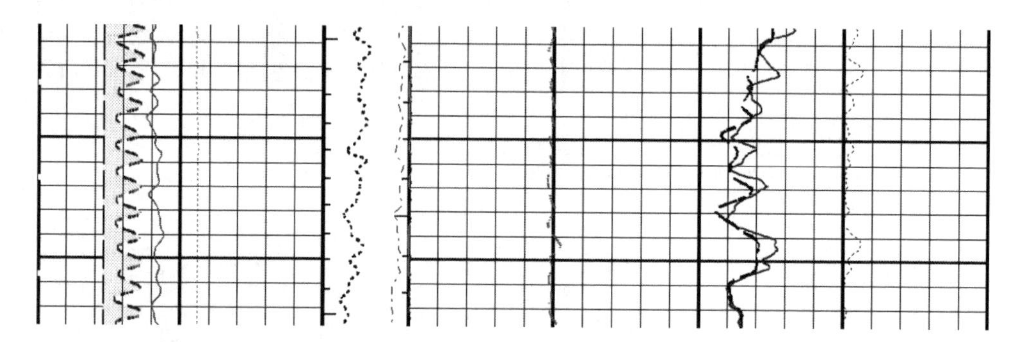

Figure 9.6

Pips shown on an interval in the so-called depth track. The caliper curve (looking like a sine wave) indicates the spiral shape.

Integrated Hole/Cement Volume Summary

Hole Volume = 58.07 M^3
Cement Volume = 17.37 M^3 (assuming 10.75 IN casing O.D.)
Computed from 2104.9 M to 1410.0 M using data channel(s) HCAL.

Figure 9.7

Result of the hole volume integration. The numbers may be inaccurate if the assumptions of the calculation are not met.

The computation is only correct if the following assumptions are met:

- The hole is circular (for a one-arm caliper) or elliptical (two-arm caliper).
- The caliper arms are never fully open over the whole logging interval. If the caliper curve flattens at high values, this means that the borehole is larger than the maximum opening of the arm. In that case the borehole volume integration is inaccurate.

In Fig. 9.6, the hole has a spiral shape. The shape assumption is not met. In Fig. 9.8, the calipers open to their maximum. Again the assumption required by the calculation is not satisfied.

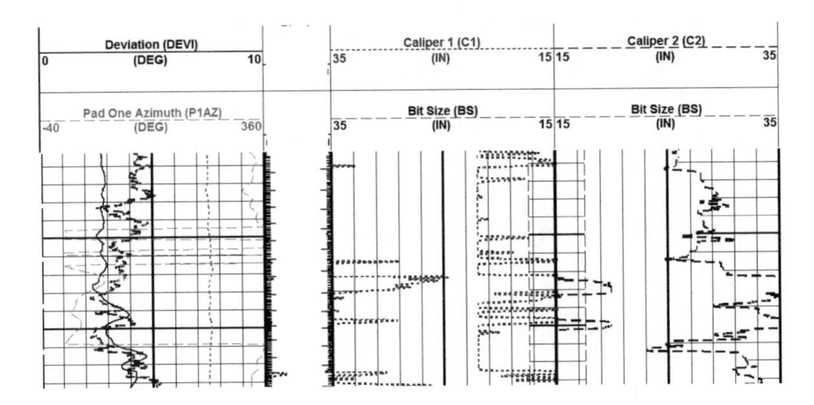

Figure 9.8

Dual caliper tool showing very large values. The dotted curve in track 2 displays a flattening at 42 in. This means that the tool is fully open and that the borehole is larger than 42 in. The dashed curve in track 3 shows a short flat shape at 40 in. Two remarks apply: 1/The tool specification is such that the arms open to a maximum of 40 in. The reading at 42 in demonstrates that the tool is not properly calibrated. 2/The curves shown are "backups" corresponding to readings above 35 in.

9.5 CALIBRATIONS

In the last twenty years, calibration information has been shown as graphical tails. Calibration tails do not give complete information on the real status of the calibration. The "OK" flags comfort the data user that everything is under control.

In the example shown on Fig. 9.9, all calibration flags are displayed as acceptable. In reality, the tool is not properly calibrated because the calibration area is not following the logging company standards. The tool is 8 mmhos off, corresponding to a 40% error for a true resistivity of 50 ohm-m. In conclusion, the calibration tail is not conclusive to verify that the calibration is correct.

The calibration process includes many tasks. These tasks involve pieces of equipment that are not easily understood except by the logging engineer or the logging technician. Logging companies should volunteer the documentation that explains calibration setups, auxiliary equipment and operating procedures. One data vendor has produced such documentation [1].

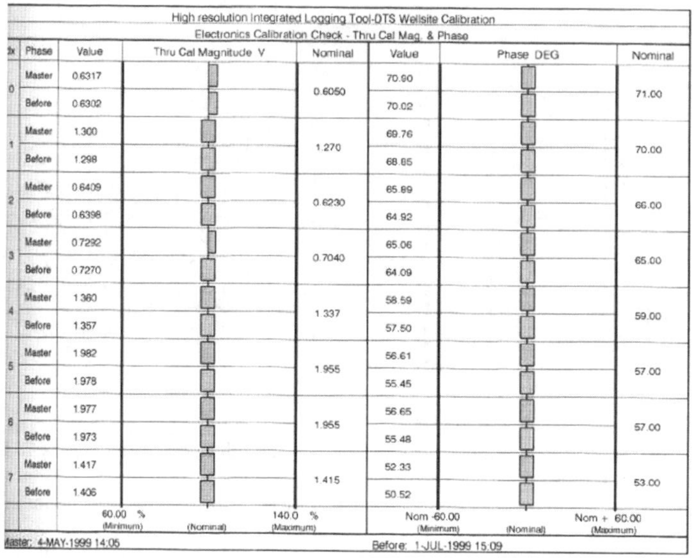

High resolution Integrated Logging Tool-DTS Wellsite Calibration
Electronics Calibration Check - Thru Cal Mag. & Phase

Idx	Phase	Value	Thru Cal Magnitude V	Nominal	Value	Phase DEG	Nominal
0	Master	0.6317		0.6050	70.90		71.00
	Before	0.6302			70.02		
1	Master	1.300		1.270	69.76		70.00
	Before	1.298			68.85		
2	Master	0.6409		0.6230	65.89		66.00
	Before	0.6398			64.92		
3	Master	0.7292		0.7040	65.06		65.00
	Before	0.7270			64.09		
4	Master	1.360		1.337	58.59		59.00
	Before	1.357			57.50		
5	Master	1.982		1.955	56.61		57.00
	Before	1.978			55.45		
6	Master	1.977		1.955	56.65		57.00
	Before	1.973			55.48		
7	Master	1.417		1.415	52.33		53.00
	Before	1.406			50.52		
			60.00 % (Minimum) (Nominal) 140.0 % (Maximum)			Nom -60.00 (Minimum) (Nominal) Nom + 60.00 (Maximum)	

Master: 4-MAY-1999 14:05 Before: 1-JUL-1999 15:09

Figure 9.9

Calibration tail with all calibration coefficients within tolerance.
Still the calibration is not correct because the calibration setup is not adequate.

9.5.1 Caliper calibration case study

To explain that calibration acceptance flags (green, yellow, etc.) are not a definite guarantee that calibrations are valid, the example of the caliper calibration is taken.

Calipers are generally calibrated in rings of specific internal diameters (ID). The caliper arms are fully open in the rings. Raw values are recorded while the arms are in these rings. As the real ID values are known, it is easy to compute the gain and offset required to obtain correct readings. For a correct calibration with rings in good condition, the following values are observed.

- The caliper raw value is 7.019 when the arms open in a 8-in ring.
- The caliper raw value is 10.36 in a 12-in ring.

From these two tasks, it is possible to derive gain and offset so that:

- Calibrated value = raw value * 1.197 – 0.40

Sometimes, proper rings are not available for a logging job. Cut-off rings as depicted on Fig. 9.11 are used. Because of the cut, the rings are deformed during transportation and storage. The small ring has an actual usable diameter of 8.5 in. The large ring has a diameter of 11.5 in. The raw values in these rings are respectively 7.437 and 9.942. The calibration performed with these rings looks like Fig. 9.12.

The gain and offset for this new calibration are different. From the same raw values, new calibrated values are computed. Table 9.2 indicates what the caliper curve displays for a range of values. At 16 in the error is 2 in. Is it important? The caliper is used to quantify the

High resolution integrated logging tool-DTS wellsite calibration					
HILT caliper calibration					
Phase	HILT Caliper Zero Measurement IN	Value	Phase	HILT caliper plus measurement IN	Value
Before		7.019	Before		10.36
6.000 (Minimum)	8.000 (Nominal)	10.00 (Maximum)	9.000 (Minimum)	12.00 (Nominal)	15.00 (Maximum)
Before: 30-JUN-1999 17:31					

Figure 9.10

Example of calibration. The flags look good. The rings are complete and in good shape.

Figure 9.11

Cut-off calipers with 8.5-in and 11.5-in internal diameters.

High resolution integrated logging tool-DTS wellsite calibration					
HILT caliper calibration					
Phase	HILT Caliper Zero Measurement IN	Value	Phase	HILT caliper plus measurement IN	Value
Before		7.437	Before		9.942
6.000 (Minimum)	8.000 (Nominal)	10.00 (Maximum)	9.000 (Minimum)	12.00 (Nominal)	15.00 (Maximum)
Before: 30-JUN-1999 17:31					

Figure 9.12

Tail of the calibration performed with cut-off rings. The rectangular box on the left is shortened. The rectangular box on the right is longer. The two boxes are still within the limits tolerated by the logging company.

volume of cement to be used and to correct the borehole effect of the induction resistivity and neutron porosity curves. Errors of several porosity units (pu) result on the neutron porosity.

Table 9.1 Impact of inferior calibration on neutron porosity

Actual size	Raw caliper value	Log #1 with correct calibration	Log #2 with incorrect calibration	$\Delta\phi_n$
in		in	in	pu
8	7.019	8	7.333	– 1.000
8.5	7.437	8.5	8.000	– 0.750
10	8.690	10	10.000	0.000
11.5	9.942	11.5	12.001	0.751
12	10.360	12	12.667	1.001
14	12.031	14	15.334	2.002
16	13.701	16	18.001	3.002
18	15.372	18	20.668	4.003
20	17.042	20	23.336	5.003

9.6 REMARKS

Remarks may not be reflected by the actual processing of the data. In general the information in the lists of parameters is more correct than the one found elsewhere. Fig. 9.13 shows the remark in the "Remarks section" contradicting what is reported in the parameter summary.

Remarks section

→ 5. Spectral Gamma Ray corrected for hole size and potassium.

6. Maximum recorded temperature is 166 DegF from temperature sensor in logging head.

Parameter summary listing

HNGS-BA: Hostile Natural Gamma Ray Sonde

BAR1	HNGS Detector 1 Barite Constant	0.981705
BAR2	HNGS Detector 2 Barite Constant	0.964037
BHK	HNGS Borehole Potassium Correction Concentration	0 ←
BHS	Borehole Status	OPEN

Figure 9.13

The top part of the picture is extracted from the remarks section. It reports that the presence of potassium in the mud is compensated. In the "Parameter summary" section, the correction is denied. The parameters in the summary are what counts and what acts on the processing, regardless of what the remarks may express.

9.7 TOOL SKETCH

The tool sketch entries are independent from what is used by the processing. On Fig. 9.14, a 2.5-in standoff is reported on the tool sketch while the processing, using the value of the parameter summary listing (Fig. 9.15) is based on a standoff of 1.5 in. The impact of an inaccurately reported standoff can be quantified. Refer to [2].

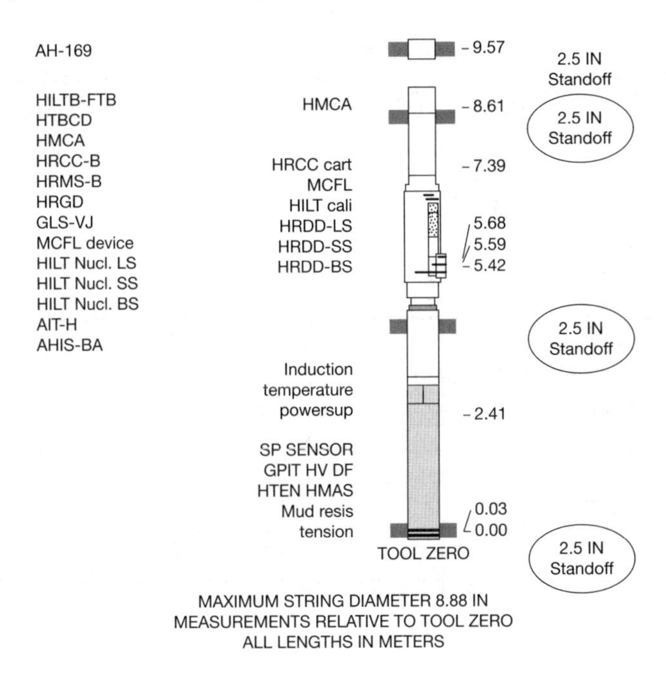

Figure 9.14

2.5- in. standoffs are reported in the tool sketch. Note that the serial numbers have been removed by the author. A proper tool sketch should include serial numbers.

Parameters

DLIS name	Description	Value
AHBHM	AIT-H Bhole correction mode	2_Compute Standoff
AHBLM	AIT-H Basic logs mode	6_One_two_and_four
AHBPO	AIT-H Basic logs processing option	Standard_processing
AHCDE	AIT-H Casing detection enable	No
AHCEN	AIT-H Tool centering flag (in borehole)	Eccentered
AHCSED	AIT-H Casing shoe estimated depth	- 50,000 M
AHMRF	AIT-H Mud resistivity factor	1
AHSTA	AIT-H Tool standoff	1.5 IN

Figure 9.15

On the same print, the parameter summary listing indicates that the standoff size (AHSTA) is actually 1.5 in.

9.8 VALIDATED AND COMPLETE INFORMATION

Most of the ambiguities displayed by data and shown in the previous sections are related to the graphical files. The presentations derived from the digital data can also be misleading. Fig. 9.16 represents the screen built from the LAS deliverable of a density log. The curves give an apparent feeling of robustness.

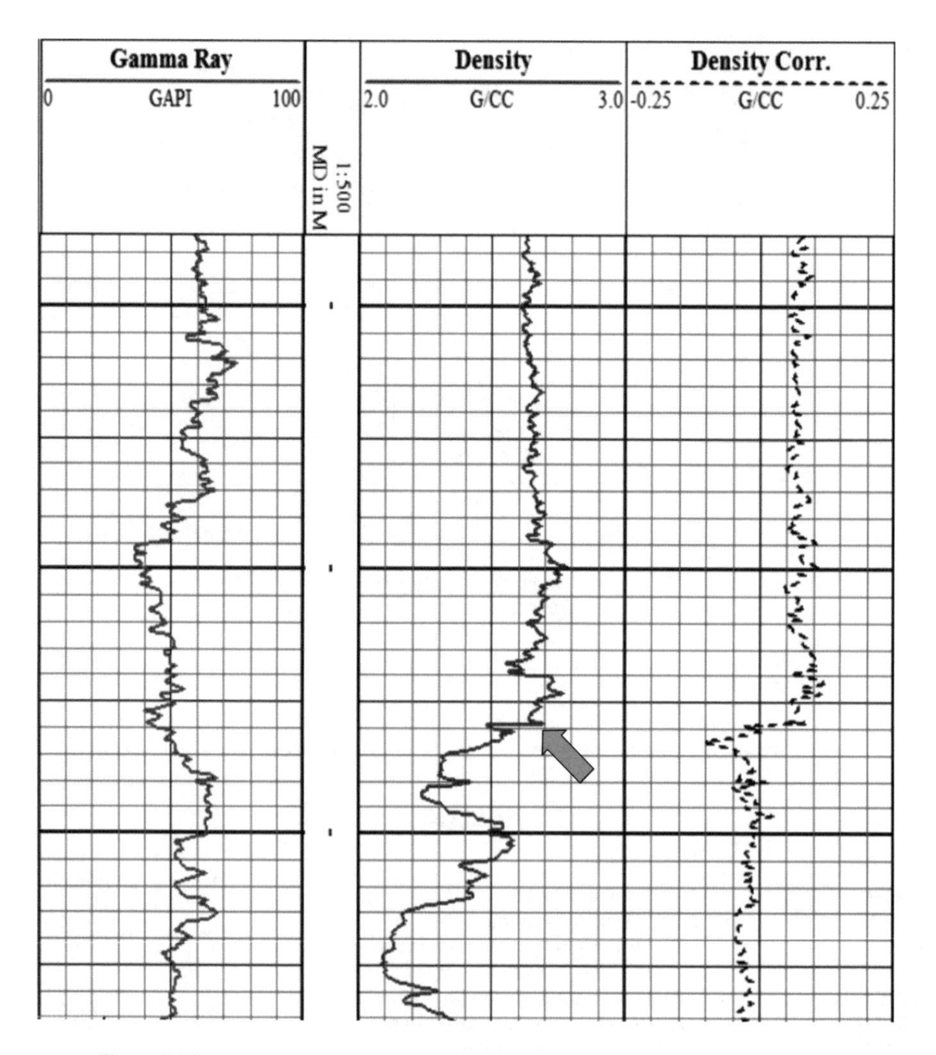

Figure 9.16

Plot obtained by loading LAS data in RIS-View (the software was used, by courtesy of the Digital Formation company).

The anomaly highlighted by the arrow (not part of the original plot) cannot be explained unless the graphical deliverable (print) is available.

Fig. 9.17 displays a graphical file recorded in the same oilfield. It contains small details that help make a better use of the data set. On the left side of track 2, it is possible to observe a short text: it indicates that there is change of bit run. On the right side of track 3, there is a black long line. This information could not be observed in the previous figure.

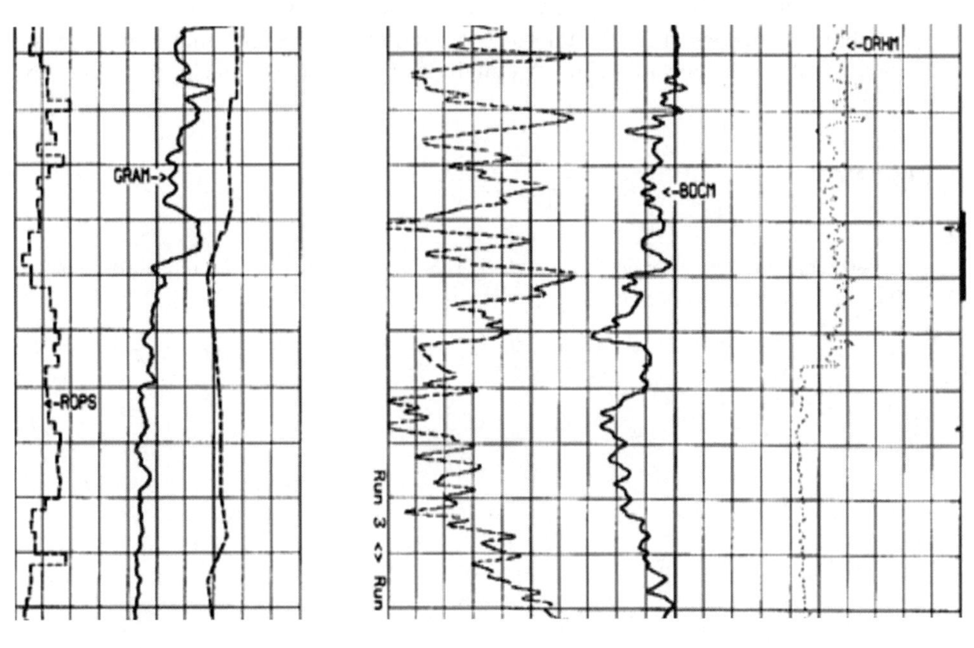

Figure 9.17

Graphical deliverable.

9.8.1 Annotations

Similarly, the display derived from the digital record does not include the annotations generated for the graphical file. Fig. 9.18 is a separate example. The annotation text reports that the information is not quantitative as the calibration is not available. Such a comment is critical to the data user. It is not available in the digital file (in LAS or DLIS format).

NOTE: data not quantitatively valid (source lost in hole therefore no tool's recalibration possible).

Figure 9.18

Annotation seen on a graphical file, but not present on a digital file.

9.9 SUMMARY

- Graphical presentations of logs are often ambiguous.
- The user needs to verify what are the numerous assumptions used to present information.
- Remarks and sketches are required to be cross-checked with other information.
- Calibration tails do not tell the full story. Even if the calibrating equipment is at fault, calibration coefficients may show within the limits accepted by the logging company.

REFERENCES

[1] Schlumberger marketing services, *Schlumberger logging calibration guide*, 08-FE-014, 2008.
[2] Theys, P., *Log data acquisition and quality control*, Éditions Technip, 1999.

10

Misconceptions

Before embarking in the review of the solutions to improve well data quality, it is necessary to list a number of behaviors and preconceived ideas that may affect progress in the quest for better quality data.

10.1 THE SUPREMACY OF REAL TIME AND OF SHORT TERM

When given the choice of two projects, one giving a return of 3$ in three years or 5$ in ten years for the same initial 1$ investment (the return dollars are expressed in present day value), many people would go for the short term project even though the return is significantly higher for the second option. This attitude results in the following decisions:

- Many companies were burning natural gas off at the wellhead at the beginning of the XXth century, because tapping this valuable resource implied additional equipment and investment.
- Field recovery is not optimized because it often requires lower production rates and hence generates cash flow more slowly.
- Data is sometimes seen as a disposable item, used for its real time value. Its long term usage is poorly understood. For this reason, proper archiving and checking are not managed in the best conditions. The cost of archiving is often considered as excessive by management.
- The short term goal of drilling faster often wins in opposition to the longer objective of getting optimal data.

10.2 NORMALIZATION

It is often thought that, even if the data from many wells in the same field is confusing, normalization will fix all issues.

Large oilfields are often surveyed by several data acquisition companies, with a large number of logging tools run by many field engineers. The characteristics of the same geological horizon vary much because of the lack of stability of the data acquisition process. Values can be found all over. Many geoscientists believe that the chaos and inconsistencies can be sorted out by normalization [1].

10.2.1 Example of normalization

- A field has 600 wells. A log measuring porosity is considered.
- 400 logs are correct and read 23 pu.
- 80 logs have been run with a calibration using inferior calibration equipment. The log reads 24.5 pu.
- 120 logs have no correction for standoff and read 26 pu.
- The porosity is averaged over the 600 wells. The average porosity is 23.8 pu.
- After normalization, 600 wells are inaccurate, instead of 200.

This example is a caricature and many petrophysicists perform normalization with utmost care. Normalization may be the only practical way to perform a field study, especially when a tight deadline is imposed.

Still normalization should be only undertaken when every other approach has failed. In that case the following process should be used:

- Zones that are geologically stable and laterally extent are recognized and separated.

Logs are quality-controlled. Inferior logs are excluded from normalization.

- Recently logged wells with extensive programs are given additional weight in the normalization.
- Suspicious logs are matched to superior logs in stable zones through a transform that makes sense in terms of tools physics.
- The same transform is applied in non-stable (i.e., with poor lateral continuity) zones. Changes of well condition (borehole, mud, trajectory, etc.) between stable zones and other zones are carefully analyzed.
- All normalization steps are traced so that changes in data can be removed and eventually performed again with better control.
- It is necessary to keep in mind that surveying individual wells is needed because lack of lateral continuity does exist in Nature as can be verified by the observation of outcrops. On the contrary, normalization is based on the assumption that continuity is excellent over large distances.

10.3 DATA CLEANSING

Cleaning data sets is advertised as the ultimate solution to data problems. Data cleansing or data scrubbing is the act of detecting and correcting (or removing) corrupt or inaccurate records. Used mainly in databases, the term refers to identifying incomplete, incorrect, inaccurate, irrelevant, etc. parts of the data and then replacing, modifying or deleting this "dirty" data. [1]

1. This striking definition is borrowed from Wikipedia.

Claims on cleansing are hyperbolic:

ABC Exploration and Production of XYZ uses LMN software to ***automatically*** *clean up data to 4-6 sigma levels (quality level of 99+%).*

How can data originally uncertain by a few percents become cleaned up to a few ppm uncertainty is a mystery to most. The **automatic** attribute is all the more intriguing as perfecting a data set is hard brain work and cannot be performed by a robot or by software.

Typically, cleansing detects multiple entries (the same well entered with different names in a database). From several records, a single one is retained as a bona fide one. The other records are put to the waste basket. Cleansing is generally an automated operation based on a number of logical rules (example for an insurance company: a person needs to be male or female; he – or she – cannot be both or neither). Such logical rules may introduce a bias in the resulting data set [2].

The principle: "get it right the first time" applies well to data. Any subsequent operation may be poorly traced. It has been seen that data vendors have difficulty to freeze the number of data objects that are delivered (for a given well parameter). So, it is difficult for an external organization not expert in data acquisition to validate the completeness of a data set. Data cleansing reduces a data set, but does no add any information.

10.4 INTERPRETATION AS DATA QUALITY CONTROL

Considering that acquisition is a process difficult to control, data users often rely on log interpretation to validate well data. They rely on the strange paradigm that good data can be interpreted, while bad data cannot.

10.4.1 Example 1 of erroneous data that can be interpreted

A density tool is used over many runs in a tight and abrasive formation. The pad is getting worn. Hence, there is less metal in front of the density detectors, and more gamma rays reach the detectors. All density readings are shifted to lower values. Consequently, the density-derived porosity is reading too high. The resulting interpretation concludes in higher porosities and higher volumes of hydrocarbons, which pleases management.

10.4.2 Example 2 of erroneous data that can be interpreted

Another density tool is miscalibrated. All densities are too high. The description of cuttings indicates a few granules of pyrite. The interpreter includes the presence of pyrite in his mineralogical model. Computed pyrite content, V_{pyrite}, is too high as compared to reality, but enables a coherent interpretation.

In conclusion, coherent (but incorrect) log interpretation can be completed from inferior data sets.

10.5 THE RECOVERY FACTOR

The recovery factor is the ratio of recoverable oil reserves to the oil in place.

Two non-oilfield anecdotes may put the current recovery factor in perspective:

- A doctor explains to a healthy patient that he has a potential life span of 100 years. If he does not make a checkup every year he is expected to live 20 years. If he takes one, he may live till he is 35.
- Similarly a person is explained that his great-great-grand-parents left him 100,000 dollars, many years ago. To the best of the banking abilities and the use of very astute financial schemes, he may hope to get 35,000 dollars.

These two persons may not be very impressed. In the oil industry still, this is what is happening. A few years ago, recovery factors of 20-25% were the norm. Now 35 (%) seems to be a fair number. [2] This number is not good enough. [3]

It is obvious that a drastic improvement of the recovery factor is the high priority of the oil industry. Quality oilfield data is the cornerstone of this potential improvement. By being better informed on oil and gas fields, it is possible to have a deeper understanding of the reservoirs and produce more hydrocarbons.

10.6 SUMMARY

- Wise management integrates the values of short term and long term in spite of the natural human trend to favor short term.
- Normalization requires caution and expertise.
- Cleansing is about reducing databases, not adding to them.
- Data that can be interpreted is either good or bad. Coherent interpretation is not a test of correct acquired data quality.
- The increase of the recovery factor is the #1 challenge of the oil industry. It is helped by quality data.

REFERENCES

[1] Kane J.A., Jennings Jr., J.W., "A method to normalize log data by calibration to large-scale data trends," paper SPE 96081-MS, SPE annual technical conference and exhibition, Dallas, Texas, 2005.

[2] Muller H., Freytag J., *Problems, Methods, and Challenges in Comprehensive Data Cleansing*, Humboldt-Universität zu Berlin, Germany, 2003.

2. Some fields are claimed to have 50% recovery factors.
3. It is understood that physics preclude extremely high recovery factors.

Part 2

Quest for quality data

11

The different uses of logging data

The object of taking data is to provide a basis for action.
W. Edwards Deming, 1938

One accurate measurement is worth a thousand expert opinions.
Grace Hopper (1906-1992)

The chapters of Part 1 described the possible simplifications and misconceptions made by users about logging data. While logging programs are often similar, the information content can be used for different purposes. A strategy to optimize data acquisition starts with a better understanding of the use that will be made of the data in the future.

Edwards Deming, the ultimate quality guru, insisted that data was mainly acquired to take decisions. The corollary of this statement is that, if no decision is going to be taken, there is no real need to collect data. If the fate of a log is to be filed in an archive without being quality-controlled, looked at, or used to take decisions, there is no much point to acquire it. [1] Before the different uses of data are listed and explained, it is necessary to study the relation between data and decisions.

11.1 DATA AND DECISIONS

The decision-making process is well described in references [1], [5] and [8]. It is summarized in this section. Three important inputs intervene in this process: [2]

- the decision threshold (T),

1. Not taking action after analyzing data is also a decision. This is what the healthy person does after a medical checkup.
2. Astrology and augurs are not included in this discussion. Objective decision-taking is considered here.

- the measurement of an important parameter that describes the issue at hand, M,
- the uncertainty of the measurement, σ, resulting in an interval of confidence [M – σ to M + σ] around the measurement M.

Starting from the measurement, the lower and upper limits (V_{min} = M – σ and V_{max} = M + σ) of the interval of confidence are calculated taking into account all sources of uncertainty, following the guidelines and suggestions of references [3] and [7]. If the lower limit of the interval of confidence, V_{min}, is above the decision threshold value, T, decision D is taken. If the upper limit, V_{max}, is below T, the opposite decision, opp-D is taken.

11.1.1 Medical example: control of cholesterol content in blood

The simple and common issue of controlling cholesterol levels in human blood is considered. Table 11.1 gathers the results in four different cases.

- T: 200 mg/l,
- D: Take anti-cholesterol medication,
- opp-D: Do not take medication,
- M: Measurement of cholesterol content in blood, collected through a well-described and controlled process,
- Uncertainty on measurement: ± 10 mg/l.

In cases 3 and 4, more measurements with smaller uncertainties are to be performed.

Table 11.1 Cholesterol case study

Case	Cholesterol measurement	Decision
1	M = 270 mg/l	D is taken
2	M = 150 mg/l	opp-D is taken
3	M = 205 mg/l	no decision may be taken
4	M = 195 mg/l	no decision may be taken

11.1.2 Example of decisions taken with a bathroom weighting scale

The knowledge of a person's weight has many applications:

- It could be related to dieting or health.
- It could be the selection in a weight class for boxing or wrestling.
- In some cultures, it could be to obtain an equivalent weight in a precious metal.

The different applications call for a different magnitude of the uncertainty. Accuracy of weight is particularly critical for sport classification. [3]

11.1.3 Example of decision process with geoscience data

A parameter is selected as an input to the decision-making process. A decision threshold is determined. If the parameter is found below this threshold, the well is abandoned. Otherwise the well is fully completed. In addition, the uncertainty associated to the parameter is controlled and quantified as an interval of confidence within which the real value of the parameter is likely to be.

Case#1

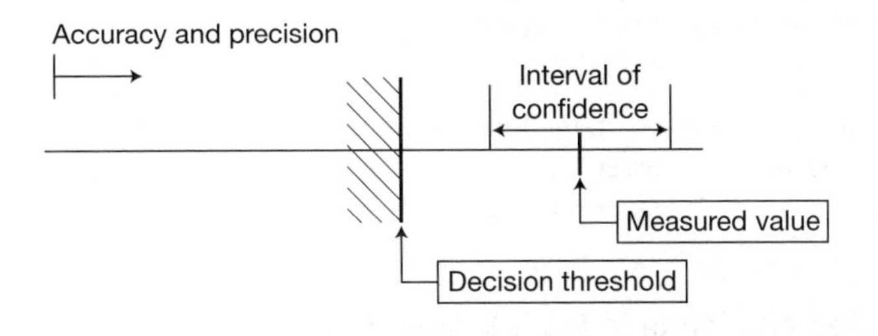

Figure 11.1

A favorable case for decision-making.

As the measured value of the parameter and its interval of confidence can be observed above the decision threshold, there is no ambiguity on the decision to be taken.

As a practical example, porosity in a shaly-sand deltaic sequence is used. The decision threshold is 20 pu. Below this value, the well is considered as dry. Above 20 pu, it is a green light for completion. The interval of confidence is ±2 pu (which, incidentally, corresponds to a mediocre quality measurement). The measured value is 26 pu which puts the real value from 24 to 28 pu, well above the decision threshold. This is a favorable case, where the task of the development team is easy.

Case#2

The measured value of the parameter is above the decision threshold, but part of the interval of confidence lies below the decision threshold.

In this second example, porosity is used in a tight carbonate environment as found in the United States Rocky mountains or in fractured formations in China. The decision threshold is 6 pu. The measurement is more accurate than in the previous example, with an uncertainty of ± 1 pu. The measured value is 6.5 pu. This means that the real porosity value is between 5.5 and 7.5 pu. The decision cannot be taken (Fig. 11.2).

3. The existence of weight divisions gives rise to intentional weight cutting. As it is an advantage to be the largest individual in a weight division, athletes try to fall in the next lower division and lose weight through dieting and dehydration prior to weighing to meet the required weight class.

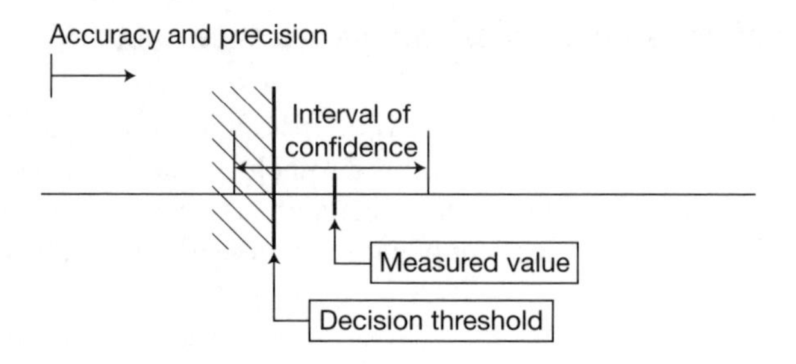

Figure 11.2

Unfavorable case in decision-making.

In the first case, the pressure on the logging company to design accurate tools and to control operating procedures is lower than in the second case where excellence is required to meet the challenging conditions imposed by the reservoir.

11.1.4 Increased challenge in log interpretation

The first example on cholesterol control looks simple enough. Conversely, it is not easy to fix decision thresholds and quantify uncertainties in petrophysics. The thresholds are often available on sophisticated parameters such as hydrocarbon volume in place, while the uncertainties can be derived, with some efforts as will be seen in Chapters 12 and 13 on basic measurements, such as density and resistivity. There are two ways to proceed:

- Starting from basic log measurements, it is possible to propagate the uncertainties to more complicated parameters. This is explained in reference [8] and depicted in Fig. 11.3.
- Starting from a complicated parameter, it is possible to define the decision thresholds imposed on basic measurements by inversion (reference [4] and Fig. 11.4).

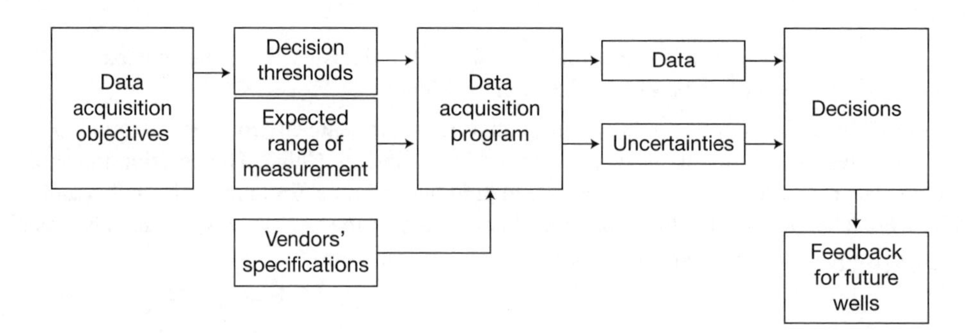

Figure 11.3

Flow chart for decision analysis.

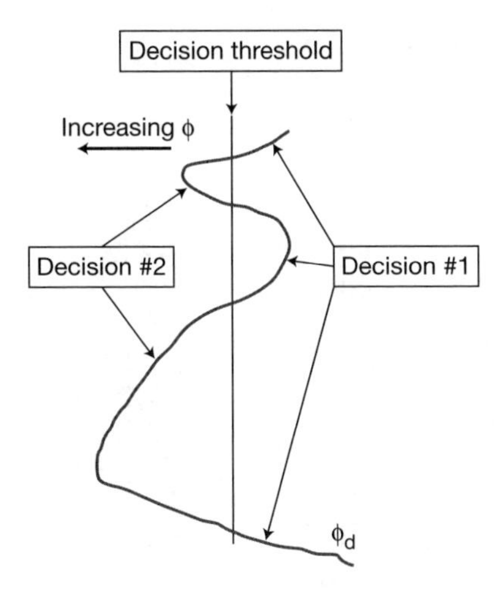

Figure 11.4

Possible scheme of decision-making with the density log.

11.2 ROLE OF LOGGING DATA

The same types of logs may be run for different purposes that are listed below in order of increasing demands on correctness (a qualitative combination of accuracy and precision).

- Make depth correlations to build the geological model,
- Perform preliminary quantification of the reserves,
- Monitor the changes of the reservoir parameters as the field is developed,
- Accurately quantify the reserves for asset determination,
- Perform unitization or re-determination.

These different roles are reviewed in the next sections. It is reminded that there are only a few opportunities to acquire data, but there are many successive uses of the data [6].

11.2.1 Depth correlations

The knowledge of a reservoir starts with the understanding of its geometry. From correlations between wells, it is possible to recognize the same horizon through a 3D-volume. Correlations can be performed with any curve showing some contrast between rock, lithology, mineralogy or fluid types. The requirements on the measured parameters are not demanding. Qualitative information is sufficient to successfully complete the correlations (Fig. 11.5).

The most important requirement for this task is to obtain precise or reproducible depth from the logging company. Fig. 11.6 shows that large errors in reservoir volumes result from poor depth control.

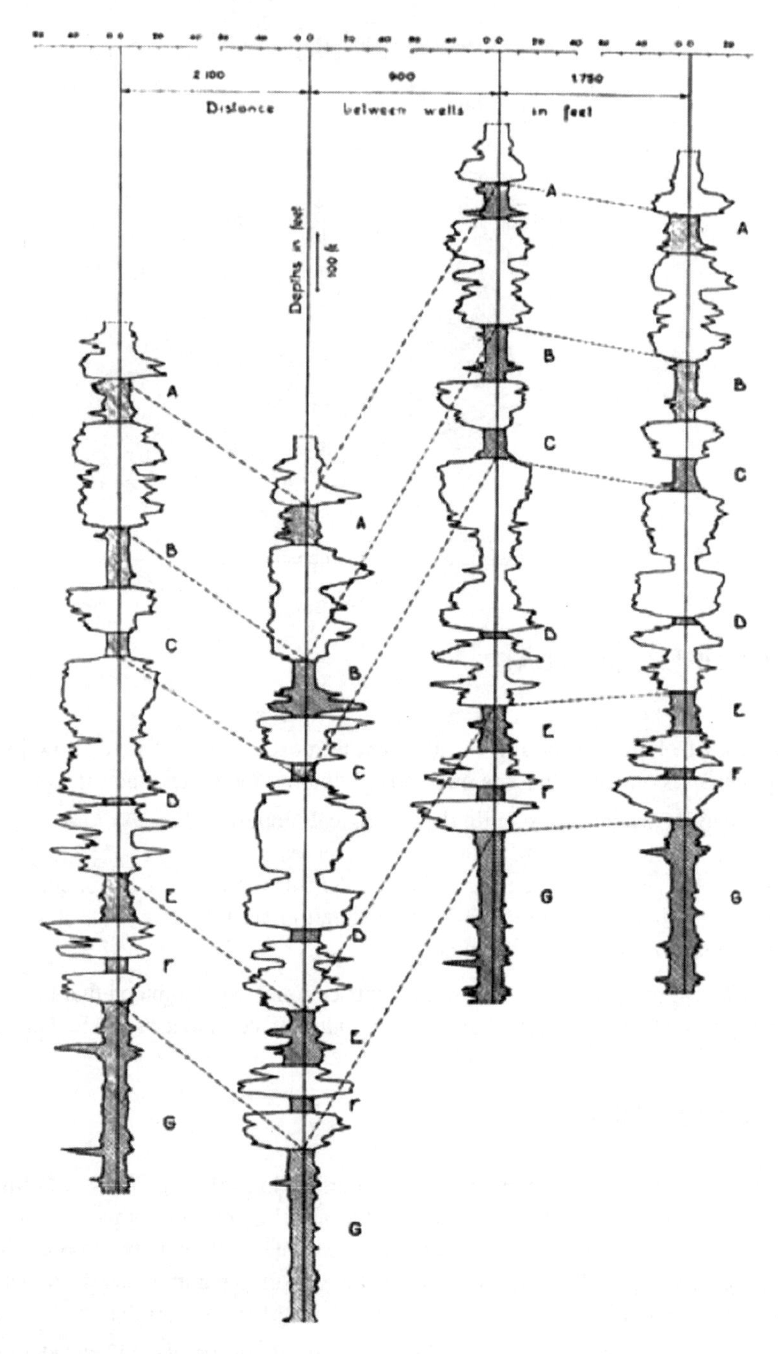

Figure 11.5

Correlation between wells.

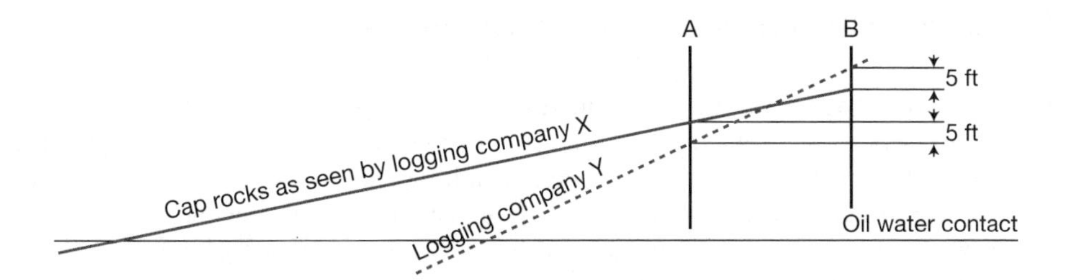

Figure 11.6

Wells A and B traverse the cap rocks. Vendor X has rigorous depth procedures and delivers correct depth. Vendor Y has inferior depth control procedures. In well A, measured depth is 5 ft deeper than correct depth; in well B, it is 5 ft more shallow, and hence the size of the reservoir is evaluated as much smaller. Note that in this specific example the oil water contact is not traversed by wells A and B.

Depth-related issues are covered in Chapter 15. It is to be noted that some formations do not lend themselves to easy correlations (Fig. 11.7).

Figure 11.7

Poorly correlatable formation (Argentina outcrops).

11.2.2 Quantitative reserve evaluation at an early field stage

Once a few wells have been drilled and the reservoir shape is understood, it is time to complete an early evaluation of the hydrocarbon volume in place. The numbers so acquired are temporary and are to be revised with any additional information. The demands on the accuracy of the petrophysical parameters are now higher as quantitative measurements are needed. The recovery factor remains one of the most poorly defined parameter. [4] At this stage, high risk situations may be encountered (Table 11.2).

Table 11.2 High risk situations

Challenges	Comments
Velocity anomalies	An isochronous anticline may be transformed into an isobathic syncline.
Flat reservoir closure	Small velocity variations change the isobathic map.
Compartmented traps	Different pressure regimes
Tilted oil water contact	Potentially different hydrodynamic regime
Undetected fluids contact	
Lateral variation of reservoir thickness	
Lateral variation of porosity	
Lateral variation of lithology	
Vugular or non-connected porosity	Affects permeability, capillary pressure and flow characteristics
Uncertainty on hydrocarbon typing	
Low porosity reservoirs	Problems of sample representativeness
Thin bed formation	
Multi-mineral formations	

11.2.3 Enhanced recovery

Previous data uses have mostly involved open-hole logging. Once the field is being produced, reservoir monitoring through cased-hole logs is necessary to optimize field development.

The same parameter (e.g., the water saturation) is measured as a function of time. The stability of the measurements and the reproducibility of the tools are required. In particular, cased-hole water saturation monitoring in carbonates, when the separation between the water line and the oil line is small, is particularly challenging.

11.2.4 Field study

After some years of production, many development wells are drilled and allow a much better knowledge of the reservoir. Field studies are undertaken by oil companies. In spite of the huge amount of data, accuracy on the results cannot always be expected.

4. The current world average recovery factor from oil fields is 30-35% (versus 20% in 1980). This parameter ranges from a 10% average in extra heavy crude oils to a 50% average in the most advanced fields in the North Sea.

On a specific example, cased-hole and normalized open-hole neutron logs were analyzed in great details for time-lapse gas-oil gravity drainage monitoring. Despite the best efforts of record-keeping and correcting the logs in the same way, a small range of uncertainty of 0.5 pu could not be obtained. Typical discrepancies were in a 2-3 pu range. Eventually all raw data count rates were investigated. Due to variations in processing and to variations in the calibration, it was not realistic to expect an overlay to better than ± 2 pu.

11.2.5 Unitization and redetermination

Quantitative field evaluation is all the more important when a field is shared between different partners [2]. A small change in the petrophysical results can be translated in very large volumes of hydrocarbons.

Example: For a given field, recoverable reserves are estimated to total 250 million barrels of oil. At a price of 80 US$ per barrel, the total field production represents 20 billions of dollars. A change of 0.001% in the share represents 200,000 US$, not a negligible amount of money. Table 11.3 represents a realistic change of share for four companies.

Table 11.3 Example of redetermination

Partners	Original share	New share	Difference	Millions $
Company A	28.00	26.9425	– 1.06	– 211.50
Company B	12.00	11.1744	– 0.83	– 165.12
Company C	36.00	34.6605	– 1.34	– 267.90
Company D	24.00	27.2226	3.22	644.52

11.2.6 Example of multiple use of data

An appraisal well is drilled to confirm the geological model, but also to generate cash flow rapidly. The well is put in production as soon as it is drilled and cased. A porosity log is considered.

- Step 1: The log is used for qualitative prediction of the geological model and the beginning of production. The decision is taken on the difference between the readings on the exploration well and on this well.
- Step 2: The residual oil volume is monitored at the beginning of a secondary recovery project.
- Step 3: The residual oil saturation is monitored at the beginning of a tertiary project.
- Step 4: The production sharing is re-determined.

Table 11.4 summarizes the accuracy/precision requirements corresponding to the different stages of the project.

Table 11.4 Data requirements versus maturity of the project

	Depth reproducibility	Data Accuracy	Data Precision
Well-to-well correlation	*****	**	**
Basic interpretation	***	***	***
Complex interpretation	***	****	****
Enhanced recovery	****	***	*****
Redetermination	*****	*****	*****
Asset determination	*****	*****	*****

11.3 VALUE OF DATA

The different uses of data have been reviewed. The particular feature of data is that it can be used successively for different purposes. The stone shown in Fig. 11.8 is 5,000 years old and originates from Mesopotamia. Through millenaries, it can be read over and over and translated. The text, a cooking recipe, can be used hundreds of times.

Figure 11.8

This data is 5,000 years old.
By courtesy of Le Louvre museum.

Similarly, well data can be used over the life of the field, which can be expressed in decades. As seen before, the porosity log can be used a first time, for a quick and quantitative assessment of reserve, a second time for a more quantitative approach, then as a reference in a time-lapse monitoring application. Finally, it can be used in redeterminations. The data life span could exceed forty years.

In most cases, the decision to collect data is made on the value of the first use. Future uses are seldom considered. In the example depicted in Fig. 11.9, it is decided to pay 100,000 US$ because the immediate value is estimated at 150,000 US$.

Table 11.5 Comparison of cost and value of data in the field life

Year	1	4	7	10	13	16	19	22	Total
Cost (k$)	100	2	2	2	2	2	2	2	142
Value (k$)	150	20	25	30	40	50	50	80	465

The same data is used at year 4 (value: 20 kUS$), 7 (value: 25 kUS$), 10 (30 kUS$), 13 (40 kUS$), 16 (50 kUS$), 19 (70 kUS$) and 22 (80 kUS$). The estimation of the value of future uses is conservative. Data used in redeterminations has often a very high value as it leverages large changes in field ownership. In the example, the cumulative value of data is 465 kUS$ in constant dollars. Meanwhile, the cost is limited to the original invoice (100 kUS$) and the payment of the archival cost (2 kUS$/year). Cumulative cost is 142 kUS$. Cumulative value is 465 kUS$. Spending the money to collect (and preserve/archive) the data is definitely a wise choice.

On the side of the data vendor, a poor data set could also be a large liability. If it is later demonstrated that insufficient efforts have been provided to collect professional data, a bad data set may result in a loss of contracts or in an erosion of future log prices.

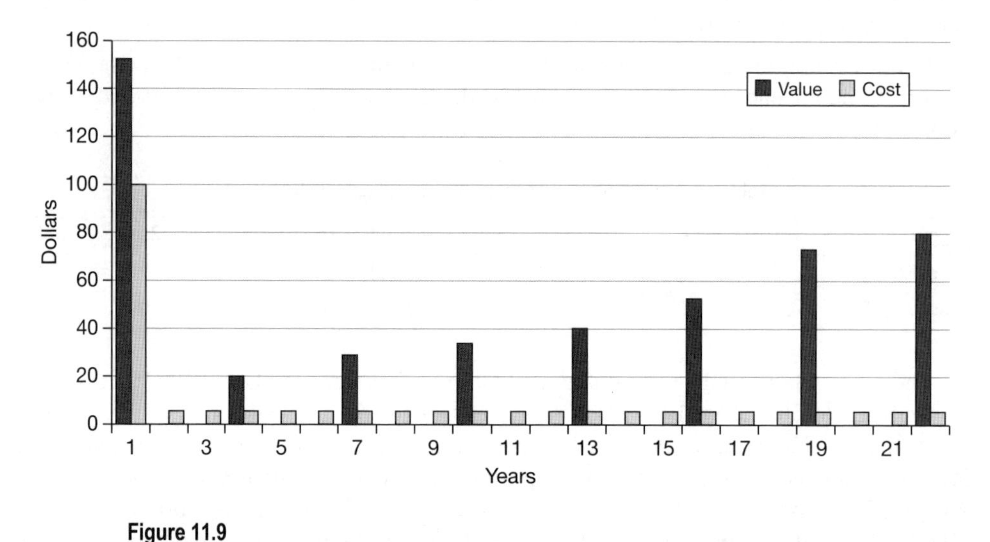

Figure 11.9

Value and cost of data in the long term.

In fact, data is the only asset that can be reused at low cost. People need to be paid again. Equipment needs to be maintained and refurbished. Data just needs to be archived in a professional manner. But data can only be reused if it is available, complete, documented, traceable and unbiased. Data to be used decades from now is difficult to validate as the future quality requirements are not yet set at the time of acquisition.

11.3.1 Invisible data loss

It may happen that shortcuts in the data acquisition process put future use in jeopardy. An example is used: twenty years after acquisition, the data is needed in a field sharing battle. It is then discovered that the calibrations were poorly completed. The data of this key well cannot be included in the negotiation. Millions of dollars may be lost.

11.4 SUMMARY

- Data is used for many different purposes.
- Data can be used for decades.
- Accuracy and precision requirements vary with each application.

– Data loss may not be immediately detected. Its financial impact may be catastrophic years after acquisition.

REFERENCES

[1] Blacker, S., "Decision-making process for streamlining environmental restoration: risk reduction using the data quality objective process," 19th annual national energy and environmental quality division conference, 1994.

[2] Davis. N., Downing. J. A., Gouldstone. F., Lolley. R., "The mathematics of unitisation-a zero sum game," SPWLA, 10th European formation evaluation symposium transactions, Aberdeen 1986.

[3] ISO, International Organization for Standardization, *Guide to the Expression of Uncertainty in Measurement (ISO/TAG4/WG3),* Genève, 1995.

[4] Kimminau, S., private communication.

[5] Liu, S., Ford, J., "Cost/benefit analysis of petrophysical data acquisition," SPWLA 49th annual logging symposium Edinburgh, Scotland, May 25-28. 2008.

[6] Louis, A., Boehm, C., Sancho, J., "Well Data Acquisition Strategies," SPE 63284.

[7] Taylor B.N., Kuyatt C. E., *Guidelines for evaluating and expressing the uncertainty of NIST measurement results.*

[8] Theys, P., *Log data acquisition and quality control*, Éditions Technip, 1999.

12

Brochure Specifications

Better than the previous measurement.
Unique in the industry.
Twice as accurate as…
Vendors' specifications documents

The need to understand the relation between

- measurement,
- decision threshold, and
- interval of confidence

was demonstrated in the previous chapter. The logging measurement is to be delivered unbiased. The decision threshold comes from the oil company management. What about the interval of confidence? As it is not an easy matter, the quest for this value is undertaken through two chapters:

- starting from the data vendor specifications,
- applying these specifications to realistic well environments, and quantifying real uncertainties.

12.1 IMPORTANCE OF SPECIFICATIONS

Surveys among petrophysicists and other data users demonstrate that log specifications are poorly known, even though this knowledge is critical. Without comparing actual measurements to the specifications offered by the vendors, it is not possible to evaluate whether a measurement is valid or not.

This is explained in Fig. 12.1. The rock sequence represented on the left is logged by three different logging tools. It includes a thick 4-ft bed and a thin 6-in bed. The first vendor comes out with the curve on the left, the second vendor with the curve in the middle, the third vendor with the curve on the right.

In fact, it is impossible to judge the validity of these logs without the additional information provided by the vendors and gathered in Table 12.1.

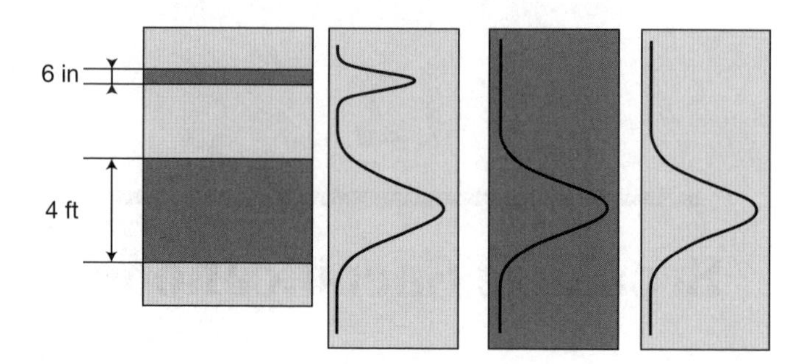

Figure 12.1

Importance of specifications. Which curve is invalid? Note that the curves in the middle and on the right look identical.

Table 12.1 Vertical resolutions claimed by different vendors

Vendor	Specified vertical resolution
Vendor 1	3 in
Vendor 2	3 in
Vendor 3	2 ft

Vendors 2 and 3 supply two curves that look the same. The curve of vendor 3 is acceptable as the tool is not designed to see beds thinner than 2 ft. But the curve of vendor 2 is not, as superior vertical resolution is claimed.

Similarly, when a vendor claims good accuracy on the measurements and it is not confirmed by the facts (whether due to calibration, environmental effects, tool failure, etc.), then the log can be classified as invalid.

12.2 SPECIFICATIONS PROPOSED BY THE DATA VENDORS

Vendors are sometimes reluctant to provide detailed specifications. They publish specifications that relate to unrealistic conditions where the measurements have a good chance to be valid. Different vendors with different tools and different technologies sometimes claim identical specifications.

12.2.1 Strong emphasis on operating specifications

Vendors tend to insist on the conditions under which the logging tool would function. They provide little information on the measurement quality and on the conditions required to obtain this level of quality. As an example, depicted on Table 12.2, pressure and temperature limits are well specified, but there is no information on the accuracy of the tool.

In this example, there is only one measurement-related piece of information, the depth of investigation. [1]

Table 12.2 Specifications for the AAX logging tool

Item	Unit	Specification
Minimum hole size	in	5 7/8
Pressure	kpsi	20
Temperature	°C	150
Length	ft	32
Sonde diameter	in	5
Total weight	lbf	1,200
Logging speed	ft/h	250 to 3,000
DOI	in	1.5 to 2.7

Little is said about how accurate the measurement is.

12.2.2 Suspiciously similar specifications from different vendors

Sometimes, specifications do not seem to have any solid ground. Logging tools have different sizes, use different detectors of different crystal composition, and are manufactured by different companies. It is therefore surprising to read that they have exactly the same accuracy specification as shown in Table 12.3.

Table 12.3 Specifications from four different vendors

Vendor	Blue	Yellow	Red	Green
Density accuracy (g/cm^3)	± 0.015	± 0.025	± 0.025	± 0.015
Density precision (g/cm^3)	± 0.006	± 0.025	± 0.015	± 0.0075

The equality of the numbers is an unrealistic coincidence.

12.2.3 Hyperbole

Specifications read in vendors' brochures tend to be exaggerated. These statements are of no use to the data user. Comparative, superlative and qualitative specifications are of little help. Hard facts are more useful. An example of an overoptimistic specification is the Schlumberger depth, claimed with an accuracy of ± 5 ft and a precision of ± 2 ft for a well at 10,000 ft. Common experience reveals that the depth accuracy in these conditions is seldom better than 10 ft. There is nothing wrong with the honest explanation of the challenges of measuring depth, but logging companies sometimes shy away from telling the oil companies about reality.

1. DOI is an ambiguous acronym: it could mean depth of investigation, diameter of investigation or even diameter of invasion.

12.2.4 Minimum specifications

In terms of metrological specifications, that is, measurement specifications, hard, correct, documented and quantitative information is expected on four attributes:

- Accuracy
- Precision
- Depth of investigation
- Vertical resolution

Auxiliary information on the field of application of these specifications is welcome and necessary.

12.3 UNDERLYING DEFINITIONS OF THE SPECIFICATIONS

There is a definite lack of agreement on the definitions used to write specifications.

12.3.1 Precision

Precision, the ceiling of random errors, is often based on the assumption that statistical errors correspond to a Gaussian distribution (Fig. 12.2).

It is worth reminding the meaning of a precision expressed with n-σ. Being given several measurements, the most probable (real) value of a formation parameter is often taken as the mean of these measurements, x, with the measurements plotting on a Gaussian curve with a standard deviation σ. σ is the uncertainty assigned to these measurements. The real value

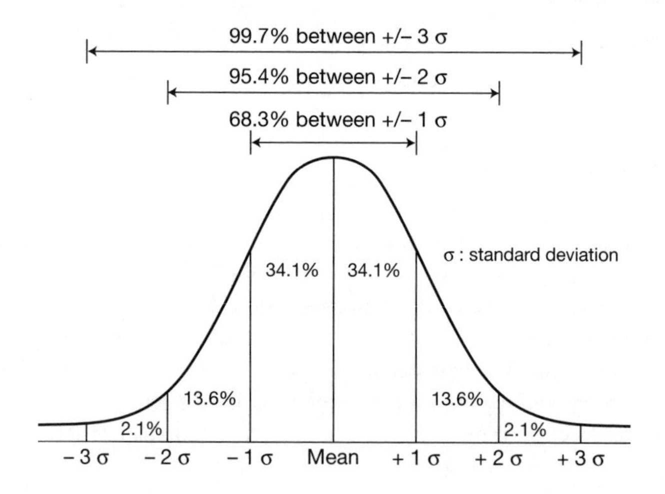

Figure 12.2

Meaning of n-σ.

has 68.3% probability to be within between $x - \sigma$ and $x + \sigma$, a 95.4% probability to be within $x - 2\sigma$ and $x + 2\sigma$ and a 99.7% probability to be between $x - 3\sigma$ and $x + 3\sigma$.

Physicists prefer the use of 2 σ, while logging companies mostly use one σ. It is recommended that the term "precision at $x\sigma$," x being defined as 1, 2 or 3, be specified.

12.3.2 Vertical resolution

There are many conflicting definitions for vertical resolution [1]. The specification should clearly define which one is selected.

12.3.3 Depth of investigation

A single value for depth of investigation is misleading. Most resistivity devices consider the value of the tool cumulative geometrical factor at 50% while 90% is taken for nuclear devices. It is preferable to list the three values of the cumulative geometrical factor, at 10%, 50% and 90% [1] and [4].

12.3.4 Conditions of applications of the specifications

The conditions under which the listed specifications apply should be listed. In general, these conditions are extremely favorable. Good specifications should show the potential impact of the well environment on the specifications. Reference to correction algorithms and correction charts would also be beneficial [3].

12.3.5 Testing the specifications in the real world

Before a logging tool is commercialized, the proposed specifications must be checked during the field testing of the prototypes or first pilot tools.

Unfortunately, there is no 3rd party organization or regulatory body that checks the specifications in an unbiased manner. Specifications figures mostly originate from the logging companies. Attempts to perform neutral testing of logging tools, such as the Spartan and Europa projects, have not been successful. After a few years of operations, they were no longer supported by oil companies. The value of these independent organizations cannot be underestimated.

12.4 GETTING ACCURACY SPECIFICATIONS

12.4.1 Quantifying systematic errors

The first step to write an accuracy specification is to list the potential (and systematic) errors that affect the measurement, [5] and [6].

They include:

- Tool response errors,
- Calibration errors,
- Environmental correction errors.

The first type, tool response error, can be easily derived from the minimization routine used to match the overall tool response equation to discrete experimental points or to the results derived from tool modeling. This error is generally quite small.

The calibration error can be experimentally obtained by getting several crews to perform calibrations on different tools.

Reasonable values for the errors caused by environmental corrections should be selected. These errors correspond to the balance left after environmental corrections are applied. Finally, systematic errors can be combined. These errors may not have a Gaussian profile and the algebraic sum may be more appropriate than the quadratic sum.

12.4.2 The challenge of improved technology

New technology is developed to satisfy the need for better accuracy. This means that the accuracy specification of the replacement of an old design is normally a smaller number (closer to zero, zero corresponding to perfect accuracy). Improving the overall accuracy only happens if each component of the accuracy budget is reduced, that is, systematic calibration errors are reduced, tool response error is reduced and errors linked to the environment are also reduced or better corrected. Fig. 12.3 shows how the reduction on the accuracy figure for a porosity tool through the use of better electronics or detectors can only be obtained with a better calibration procedure and equipment. The calibration share in the error budget is 1 pu for old technology and needs to be brought down to 0.5 pu for the new design.

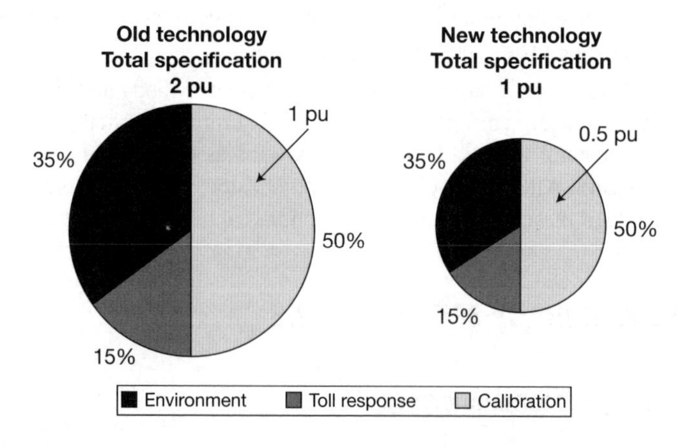

Figure 12.3

Required reduction of all contributions to the error budget for a new tool.

12.5 OBTAINING PRECISION SPECIFICATIONS

Precision mostly applies to nuclear tools (for other tools, it is generally excellent, and this is represented by a precision close to zero). It can be obtained by using the formula of error propagation applied to counting rates. It is verified by running several passes over the same interval in test wells [6].

12.6 DEVELOPING REPRODUCIBILITY SPECIFICATIONS

Reproducibility is the difference between two measurements run with different tools. This information is seldom communicated by logging companies. A rare published example is shown in Table 12.4.

Table 12.4 Reproducibility information communicated by Western Atlas

Apparent mean ϕ_n (pu)	Single instrument (3-pass σ)	Instrument-to-instrument σ
25	0.45	0.62
14	0.14	0.30
4	0.07	0.17

In this case, the reproducibility looks 50% to 100% worse than precision.
By courtesy of SPWLA.

In reality, it is easy to obtain this specification. A minimum of two tools must be run on the same well though a larger number is preferred. Reproducibility is relevant for field studies, as many wells cannot be surveyed by the same logging tool and crew.

12.7 EXAMPLES OF COMPLETE SPECIFICATIONS

12.7.1 Accuracy and precision of density measurements

Complete specifications include a split between accuracy and precision, as shown in the Table 12.5, below.

The specifications for precision cannot be used directly as they apply to densities and logging speeds that are not likely to be the ones encountered during logging in a given well. The conversion to the specific conditions of the well under consideration is reviewed in the next chapter.

Table 12.5 Density specifications for Schlumberger tools

	ADN475	LDT	Pex
Measuring range	1.7 to 3.05 g/cm^3	1.6 to 3.00 g/cm^3	1.6 to 3.00 g/cm^3
Accuracy	0.015 g/cm^3	0.010 g/cm^3	0.010 g/cm^3
Statistical repeatability	0.006@ 2.5 g/cm^3@ 200 ft/h for a 3-point averaging with a1-σ definition.	0.0038@ 2.5 g/cm^3 @ 1,800 ft/h, for a 3-point averaging with a1-σ definition.	0.0033 @2.5 g/cm^3, @1,800 ft/h, for a 3-point averaging with a 1-σ definition.

All numbers are in g/cm^3 unless another unit is used.

Example of conflict between accuracy and precision

It may happen that improved precision is linked to a serious deterioration of accuracy. For instance, The LWD density-neutron tool (type ADN) has been designed to be "driller-friendly." The precision specification looks better for a tool run without a stabilizer than with one (Table 12.6). In fact, a tool run slick would have a seriously worse accuracy.

Table 12.6 Slick tools correspond to better precision, but the accuracy is much worse.

	ADN4		ADN6	
	Slick	6 1/8 in Stabilizer	Slick	8 1/2 in Stabilizer
LS count rate (cps)	921	283	1,143	310
Precision (g/cm^3)	± 0.0032	± 0.0057	± 0.0028	± 0.0055

12.7.2 Vertical resolution of a resistivity tool

Some contractors attempt to give more than one figure for vertical resolution. It is true that it often varies with the value that is measured.

- Phase (all spacings) 0.7 ft @ 0.2 ohm-m
  2.0 ft @ 200 ohm-m
- Attenuation (all spacings) 1 ft @ 0.2 ohm-m
  8 ft @ 50 ohm-m

In this particular example, the measurement displays a worse vertical resolution at high resistivity.

12.7.3 Depth of investigation of a resistivity tool

A single number for depth of investigation is generally insufficient to understand the actual behavior of a logging tool. The figures communicated in Table 12.7 are needed by the data user to elucidate complex resistivity profiles.

Table 12.7 Depth of investigation of LWD attenuation and phase measurements

Attenuation Measurement (ohm-m)						
Geometrical factor	**Spacing**	**10 in**	**16 in**	**22 in**	**28 in**	**34 in**
radius at 1 ohm-m	10%	8.5	12.3	15.1	17.5	19.2
	50%	15.0	18.3	21.5	23.7	26.2
	90%	23.0	26.0	28.7	31.2	33.5
radius at 50 ohm-m	10%	34.7	36.1	36.2	39.1	41.5
	50%	59.8	60.7	61.6	63.7	65.5
	90%	88.2	89.0	89.8	90.8	91.7

Phase Measurement (ohm-m)						
Geometrical factor	**Spacing**	**10 in**	**16 in**	**22 in**	**28 in**	**34 in**
radius at 1 ohm-m	10%	5.4	7.0	8.3	9.5	10.6
	50%	9.1	11.2	13.2	14.7	16.1
	90%	13.6	15.8	17.6	19.6	21.1
radius at 100 ohm-m	10%	7.1	10.2	13.2	16	18.7
	50%	17.1	23.8	29.7	34.8	39.6
	90%	51.3	61.2	68.0	73.0	76.8

By courtesy of Schlumberger.

12.8 ADDITIONAL MEASUREMENT INFORMATION

Some information provided by the logging companies is not presented in the format of a specification. But it is sometimes a hidden specification that can be extremely useful to the data user.

12.8.1 Planning tables

The plot shown in Fig. 12.4 highlights the limitations of enhanced-processed induction measurements. Above 100 ohm-m, the 1-ft enhanced resolution curves are not valid. Above 500 ohm-m, the 2-ft enhanced curves cannot be used, and similarly for the 4-ft enhanced curves above 1,000 ohm-m.

The recommended standoff size table clearly indicates that the tools run with incorrect standoffs give inaccurate measurements (Table 12.8).

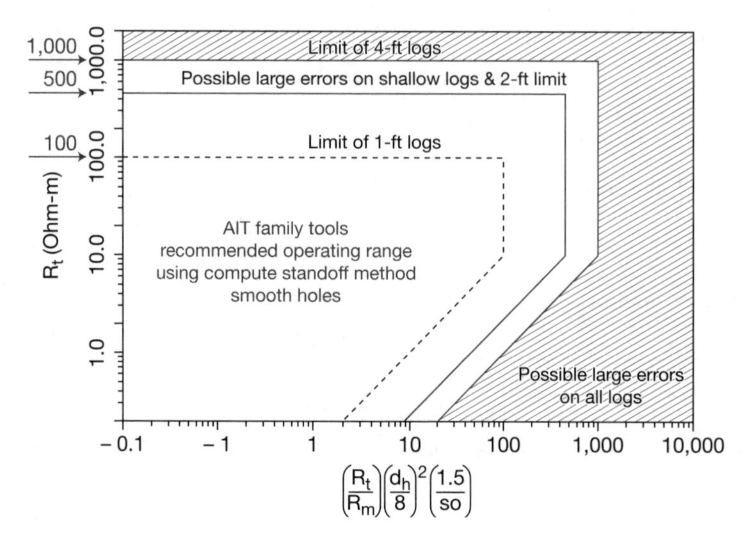

Figure 12.4

Planning of an induction logging job.
It shows the limitations of the enhanced-resolution measurements.
By courtesy of Schlumberger.

Table 12.8 Tables of recommended standoffs for AIT tools

Hole Size (in)	Recommended standoff (in)	
	B, C, H, M, HIT tool type	SAIT, QAIT
≤ 5.0	–	0.5
5.0 to 5.5	–	1.0
5.5 to 6.5	0.5	1.5
6.5 to 7.75	1.0	2.0
7.75 to 9.5	1.5	2.5
9.5 to 11.5	2.0 + bow spring	2.5
≥ 11.5	2.5 + bow spring	2.5

By courtesy of Schlumberger.

12.9 A FIRST LOOK AT UNCERTAINTIES

Brochure specifications often apply to favorable conditions. This means that, under real conditions, uncertainties will be worse. In some instances, they clearly indicate that the quest for uncertain measurements is difficult from the start. This is the case for induction tools in high resistivity environments (Table 12.9), reference [2].

Table 12.9 Specifications for an induction tool

Operating frequency	Simultaneous 10, 20 and 40 kHz
Outputs	R and X components, transmitter current
Inputs	Frequency select, tool, cal, transmitter drive
Measurement range	0.5-10,000 mS/m
R-channel accuracy	5% of reading; ± 1 mS/m
X-channel accuracy	5% of reading; ± 20 mS/m

By courtesy of Halliburton.

Accuracy is the parameter under scrutiny. 5% seems to be a reasonable number, but as soon as the resistivity goes higher than 50 ohm-m, the second term (± 1 mS/m, or millimho [2]) takes precedence (Fig. 12.5). Above this value the relative uncertainty becomes quite large. High resistivities (above 100 ohm-m) used to be laterolog [3] territory, but this type of device cannot be used when oil base mud is in the well. Logging high resistivity formation then becomes a challenge.

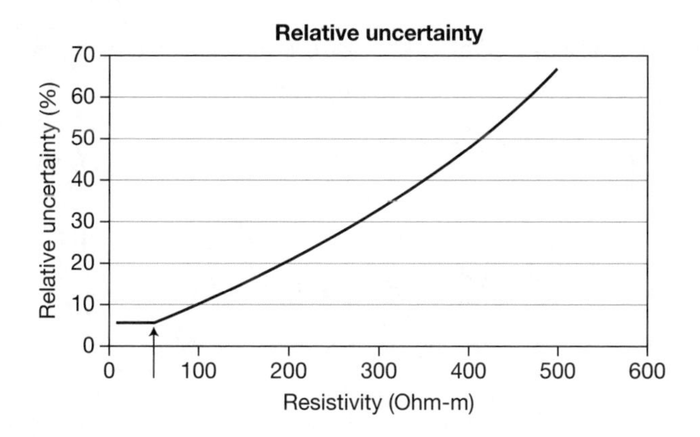

Figure 12.5

Relative uncertainty for an induction tool.

High resistivity is a challenging environment.

The arrow indicates the transition from the flat percentage number to the constant conductivity number.

12.10 PLANNING OF A LOGGING JOB WITH SPECIFICATIONS

With complete and unbiased specifications, users can decide which logging tool is suitable for their purpose. The specifications in combined accuracy and precision of three imaginary density tools are shown in Fig. 12.6. The first tool proposes ± 0.015 g/cm^3 and corresponds

2. Conductivity in millimhos x resistivity in ohm-m = 1,000.

3. A laterolog tools comfortably read resistivities, while induction tools are fit for conductive environments.

to the most expensive price tag. The two lines represent the limits of the domain where the real formation density would be. The second tool has a ± 0.030 g/cm^3 interval. It is proposed with a 30% discount. The range of values is indeed wider. Finally, a third tool is proposed with a 0.045 g/cm^3 combined accuracy and precision at a 70% discount. Depending on the use to be made of the data and the economics of the project, the data user can select between three choices. He must also take in consideration the long term use of the data.

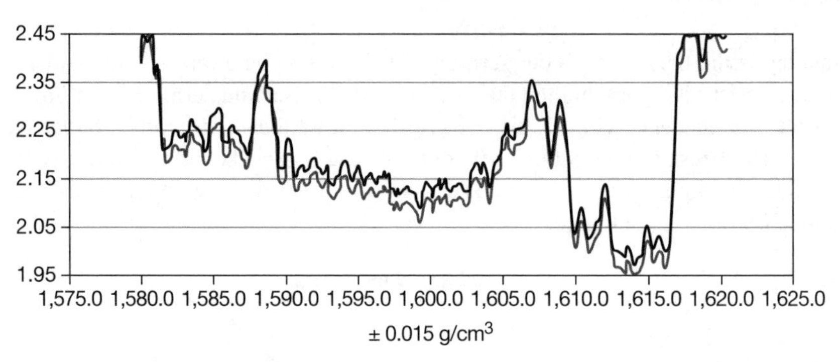

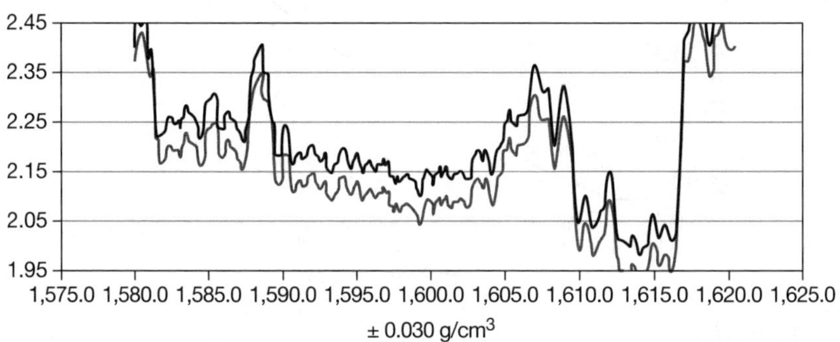

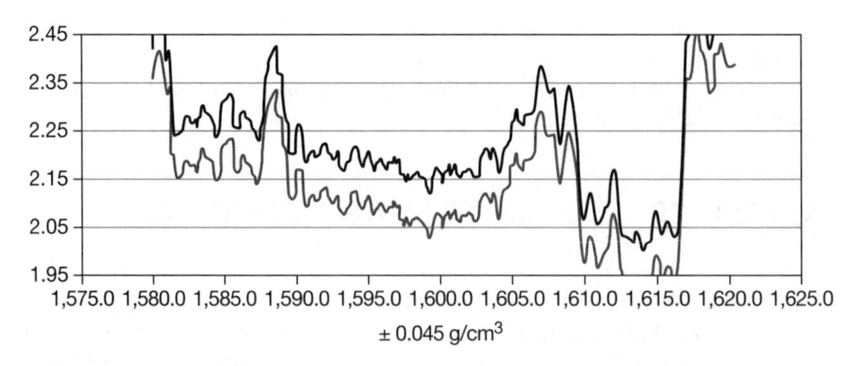

Figure 12.6

Specifications of different density tools.

The limits of the possible real values are represented by the two curves in every instance.

Table 12.10 summarizes the correspondence between measurements attributes and the type of work at hand.

Table 12.10 Different specifications are required for different conditions.

	Exploration well	Development well	Thin beds	Old wells
Accuracy	*****			
Repeatability		****		
Reproducibility		****		*****
Vertical resolution			*****	
Depth of investigation	****	****	**	****

12.11 SUMMARY

- Understanding specifications is a prerequisite to quality control.
- Current industry specifications need to be improved.
- Current vendors' specifications are not detailed enough.
- There are quantitative methods to derive robust measurement specifications. They are seldom used.
- Complete specifications can be complicated.

REFERENCES

[1] Flaum, C., Theys, P. P., "Geometrical specifications of logging tools: a need for new definitions," paper ZZ, Trans. SPWLA, 32nd annual symposium, Midland, 1991.

[2] Halliburton, http://www.halliburton.com/public/lp/contents/Data_Sheets.

[3] Schlumberger marketing services, *Schlumberger chartbook,* 2009.

[4] Theys, P., *Log Data Acquisition and Quality Control*, Éditions Technip, 1999.

[5] Theys, P., "Accuracy – Essential Information for a Log Measurement," SPWLA 38th annual logging Symposium, Houston, 1997.

[6] Theys, P., "A serious look at repeat sections," SPWLA 35th annual logging symposium, Tulsa, 1994.

13

Quest for uncertainties: From brochure specifications to real uncertainties

When it is not in our power to know what is true,
we ought to follow what is most probable.
Descartes (1596-1650)

13.1 STARTING FROM VENDORS SPECIFICATIONS

After studying the complete documentation provided by the data vendors, the data user can make a fair estimate of measurement uncertainty as

$$\text{Uncertainty}^2 = \text{Accuracy}^2 + \text{Precision}^2$$

where accuracy and precision are extracted from the vendors' specification tables. [1]

This computation leads to uncertainties that do not vary for a logging run. In the rare instances when petrophysicists consider uncertainties, [2] they use constant and reasonably small values, as shown in Table 13.1 extracted from reference [5]. [3]

---

1. The equation is valid if the uncertainties related to accuracy and precision are independent.
2. Most petrophysicists ignore uncertainties and consider that the measurements delivered by the logging company are strictly identical to the formation parameters. There is no doubt that this statement seems too strong. Any petrophysicist using a global solver (ELAN, Multi.min, Quanti.min, etc.) believes that he is dealing with uncertainties on the measurement or on the formation models or both. In the same direction, Monte Carlo methods are extensively used in the industry today. They also incorporate the notion of uncertainty. In reality, all methods use simplistic approaches on measurement uncertainty.
3. The article uses the Schlumberger tool names (LDT and CNL) but similar numbers apply to the density and neutron porosity logs delivered by other vendors.

Table 13.1 Uncertainties used in the industry

Log	GR	CNL	Rt	Sonic log	LDT
Standard deviation	± 5%	± 7%	± 10%	± 5%	± 0.015 g/cm^3

By courtesy of SPWLA.

Such an approach has several limitations. Uncertainties cannot be constant throughout a well interval as they vary with borehole conditions and tool movement, among other effects. In addition, they change with the measured value. For instance, the density precision uncertainty is larger at high densities than at low densities. [4]

13.2 REAL UNCERTAINTIES

13.2.1 Uncertainties reported by data vendors

Data vendors design and field test their logging tools. They have considerable inside information on differences between measured and real values. This is generally not available in specifications documents, but can be found in the technical literature. Two examples are taken from [1] and instruct the data user about thc magnitude of uncertainties.

Figure 13.1 relates to resistivity. Uncertainties are displayed as two thin curves on each side of the original measurement. In a high resistivity zone, the laterolog deep measurement has a small uncertainty (around 10%), while the deep induction log has large uncertainties (40%). The same article demonstrates that the resulting uncertainty on the interpreted and computed R_t is reasonably small because of the use of the two measurements collected from tools operating with different physics.

Figure 13.2 represents a density curve. Again, the two thin curves represent the lower and upper limits of the "real" value of density. The dashed curve is the original measurement. It can be noted that the upper limit is far away from the measurements over two intervals. This is explained by poor hole conditions when the measurement is biased toward too low values by mud, mud cake and rugosity. The measured density is lower than the actual formation density. It can also be observed that the theoretical density, a "best guess" of the real density, is higher than the measured density represented by the heavy curve, almost in the middle of the limits. This observation is quite common. It means that the density-derived porosity is often over-optimistic. [5]

4. Typically twice larger at 3.0 g/cm^3 than at 2.0 g/cm^3.

5. This may be one of the reasons of the mediocre commercial success of Global, the software that created the presentations shown in Figs 13.1 and 13.2.

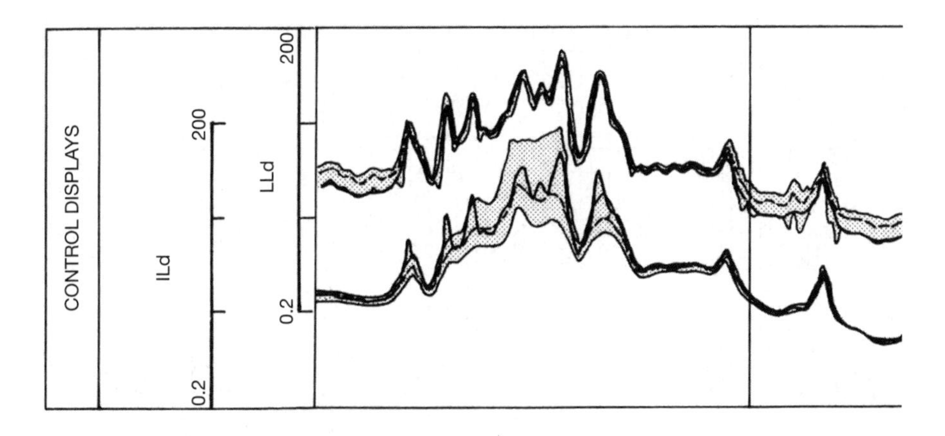

Figure 13.1

Uncertainty shown on an example of Rt Global.
By courtesy of Schlumberger.

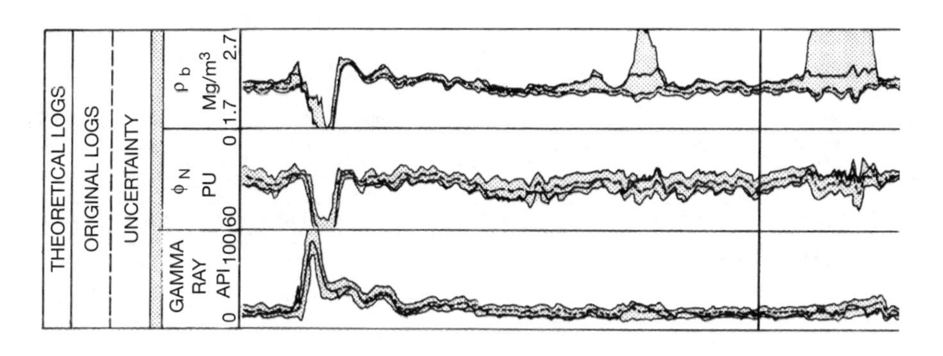

Figure 13.2

Uncertainty shown on an example of Global.
By courtesy of Schlumberger.

The two examples clearly indicate that the uncertainty does vary within the same survey.

13.2.2 Uncertainties observed when multiple passes are available

It is also possible to assess the magnitude of uncertainties by observing two measured values representing a single formation parameter. This can be obtained from two consecutive passes over the same interval or by stacking two similar tools in the logging string. In the example extracted from [3], two density log passes are acquired (Fig. 13.3). One pass is recorded in a traditional way, using classical hardware (long axis log). The second pass is derived from a logging tool with a special excentralizer (short axis logging). The differences between the two passes are listed in Table 13.2. The differences are compared to the ven-

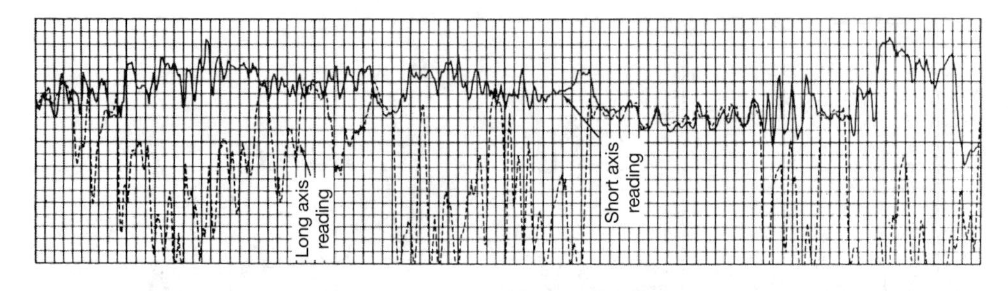

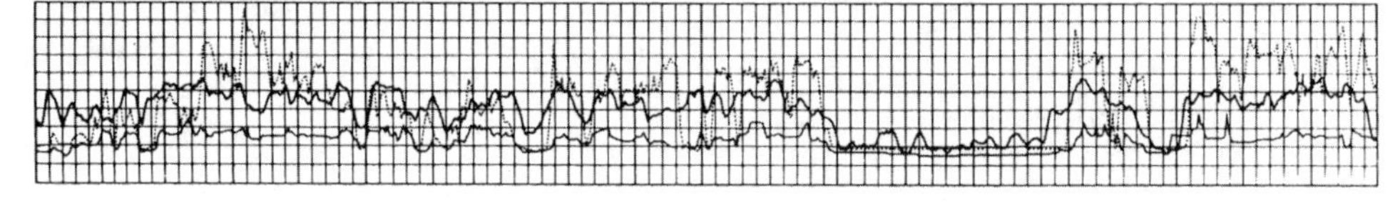

Figure 13.3

Difference between readings recorded on two separate passes.
By courtesy of SPWLA.

Table 13.2 Evaluation of uncertainties

Depth m	Reading#1 g/cm^3	Reading#2 g/cm^3	Δ g/cm^3	Δ/spec %	Δ/reading#2 %	ϕ1 pu	ϕ2 pu	Δϕ pu	Δϕ/ϕ2 %
964	2.65	2.60	– 0.05	– 333.33	– 1.92	3.51	6.43	2.92	45.45
965	2.63	2.65	0.02	133.33	0.75	4.68	3.51	– 1.17	– 33.33
966	2.60	2.60	0.00	0.00	0.00	6.43	6.43	0.00	0.00
967	2.30	2.70	0.40	2,666.67	14.81	23.98	0.58	– 23.39	– 4,000.00
968	2.40	2.70	0.30	2,000.00	11.11	18.13	0.58	– 17.54	– 3,000.00
969	2.45	2.65	0.20	1,333.33	7.55	15.20	3.51	– 11.70	– 333.33
970	2.47	2.70	0.23	1,533.33	8.52	14.04	0.58	– 13.45	– 2,300.00
971	2.65	2.65	0.00	0.00	0.00	3.51	3.51	0.00	0.00
972	2.55	2.53	– 0.02	– 133.33	– 0.79	9.36	10.53	1.17	11.11
973	2.52	2.52	0.00	0.00	0.00	11.11	11.11	0.00	0.00
974	1.90	2.55	0.65	4,333.33	25.49	47.37	9.36	– 38.01	– 406.25
975	2.07	2.70	0.63	4,200.00	23.33	37.43	0.58	– 36.84	– 6,300.00
976	2.55	2.67	0.12	800.00	4.49	9.36	2.34	– 7.02	– 300.00
977	1.90	2.70	0.80	5,333.33	29.63	47.37	0.58	– 46.78	– 8,000.00
992	1.90	2.63	0.73	4,866.67	27.76	47.37	4.68	– 42.69	– 912.50
993	1.90	2.68	0.78	5,200.00	29.10	47.37	1.75	– 45.61	– 2,600.00
994	2.60	2.70	0.10	666.67	3.70	6.43	0.58	– 5.85	– 1,000.00
995	2.55	2.55	0.00	0.00	0.00	9.36	9.36	0.00	0.00

Table 13.2 Evaluation of uncertainties *(continued)*

Depth m	Reading#1 g/cm^3	Reading#2 g/cm^3	Δ g/cm^3	Δ/spec %	Δ/reading#2 %	ϕ1 pu	ϕ2 pu	Δϕ pu	Δϕ/ϕ2 %
996	2.51	2.53	0.02	133.33	0.79	11.70	10.53	– 1.17	– 11.11
997	2.50	2.53	0.03	200.00	1.19	12.28	10.53	– 1.75	– 16.67
998	2.55	2.57	0.02	133.33	0.78	9.36	8.19	– 1.17	– 14.29
999	2.55	2.55	0.00	0.00	0.00	9.36	9.36	0.00	0.00
1,000	2.46	2.45	– 0.01	– 66.67	– 0.41	14.62	15.20	0.58	3.85

dor's specification (0.015 g/cm^3). The density readings are also converted to porosities (with a matrix density of 2.71 g/cm^3).

Out of 23 levels, only five have differences smaller or equal to the vendor's specification. They are highlighted by a grey background. The largest difference is 29.63% of the reading.

When the original readings are converted to porosities, then the percentage of the difference to the most pertinent reading is large (and up to 8,000%). This happens when the most realistic porosity is close to zero while the other reading is affected by hole rugosity.

The next sections propose a way to quantify uncertainties by following the process proposed in Fig. 13.4.

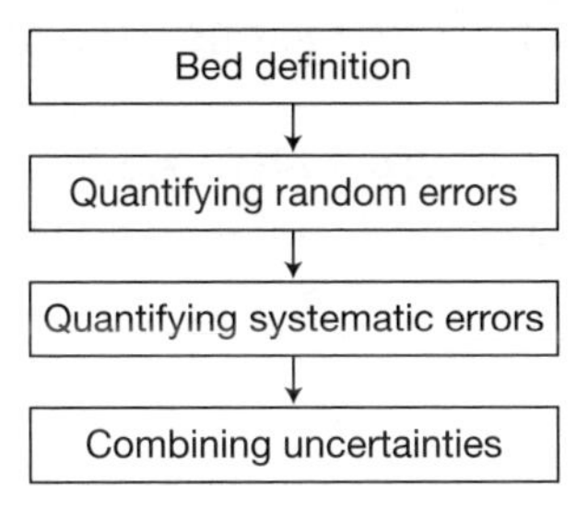

Figure 13.4

Flow chart of the quantification of uncertainties.

13.3 DEFINING HOMOGENEOUS BEDS

It is highly recommended to perform uncertainty analysis for each formation bed and not by sample, as the use of evenly sampled data (generally every 6 in) has the following drawbacks:

- Very few beds have a 6-in thickness.
- Errors due to depth matching are more visible and have a bigger impact on formations described by a small number of samples.

– Important precision improvements cannot be made if data is sampled every 6 in.

Most petrophysical software packages contain a log squaring or blocking algorithm adequate for the definition of geological units independently of sampling rates.

13.4 ESTIMATING RANDOM ERRORS FOR UNCERTAINTY ANALYSIS

13.4.1 From vendor specification to actual well conditions

In the favorable case when the vendor cares about separating precision and accuracy specifications, the precision specification is given in narrow conditions including:

- A specific logging speed or rate of penetration,
- A specific filtering scheme,
- A specific σ interval (1-σ, 2-σ or else),
- A specific measured value.

The user needs to adapt these limited conditions to the ones observed in the well under study.

13.4.2 Using a different logging speed (rate of penetration) or a different sampling rate from the one used in the vendor's specification

The precision σ_{ref} is defined by the vendor [4] [6] for a logging speed, v_{ref}, and a sampling rate, s_{ref}, expressed as a length (generally in in). If the actual logging speed, v, is:

$$v = a * v_{ref}$$

and the sampling rate, s such that:

$$s = b * s_{ref}.$$

Then:

$$\sigma = \sigma_{ref} \sqrt{a/b}.$$

This equation shows that the magnitude of σ contributed by random errors increases with higher logging speeds and higher sampling rates (when s becomes shorter than the reference sampling rate).

6. Note: a and b have different meanings in the referenced book.

13.4.3 Using a different signal processing method

If the vendor indicates in the specification document a filtering scheme different from the one used during data acquisition, it is possible to quantify the precision through the following method when the two filtering schemes are weighted-averaging.

If:

$$x_{ref} = a_1 x_{d-ps} + a_2 x_{d-(p-1)s} + \dots + a_p x_{d-s} + a_{p+1} x_d + \dots + a_{2p+1} x_{d+ps}$$

Then

$$\sigma^2_{xref} = (a_1^2 + a_2^2 + a_3^2 + \dots + a^2_{2p+1}) \times \sigma^2_{xd}$$

where the subscript d is the depth index for measurement x and s is the sampling rate.

If x, the measurement is filtered in a different way such that:

$$x = b_1 x_{d-qs} + b_2 x_{d-(q-1)s} + \dots + b_q x_{d-s} + b_{q+1} x_d + \dots + b_{2q+1} x_{d+qs}$$

Then:

$$\sigma^2_x = (b_1^2 + b_2^2 + b_3^2 + \dots + b^2_{2q+1}) \times \sigma^2_{xd}$$

Therefore:

$$\sigma^2_x = [(b_1^2 + b_2^2 + b_3^2 + \dots + b^2_{2q+1}) / (a_1^2 + a_2^2 + a_3^2 + \dots + a^2_{2p+1})] * \sigma^2_{xref}$$

13.4.4 Using a different σ

If the vendor gives a precision σ_{ref} with n-σ for value x_{ref}, what is the precision for value x, required with p-σ?

$$\sigma_x = [p/n] * \sigma_{xref}$$

13.4.5 Using a different value than the reference given by the vendor

The vendor gives precision σ_{ref} for value x_{ref}. What is the precision for value x?

A simple approach

The vendor may give a simple equation relating the precision and the measurement. Most of the time, it is of the form:

$$\sigma_x = m * x$$

This means that the precision is given as a percentage of the measurement. Then:

$$\sigma_{ref} = m * x_{ref}$$

and

$$\sigma_x = \sigma_{ref} * [x/x_{ref}]$$

Information on the measurement algorithm

If the data user has access to the simplified algorithm that links raw data to useable information, he would be able to compute the uncertainty propagated from the raw data uncertainties. If $u_1, u_2, u_3, \ldots u_p$ are the inputs and y the output function such that:

$$y = f(u_1, u_2, u_3, \ldots u_p)$$

Then the uncertainty on y is:

$$\sigma_y^2 = \Sigma_{j=1}{}^p \, \sigma_{uj}^2 \, (\partial y/\partial u_j)^2$$

It is then possible to derive the precision for any value of the measurement.

13.4.6 Example of computation of the precision for the density

Vendor specification

The MDT [7] density has a statistical repeatability of 0.0038 g/cm^3 for a value of 2.5 g/cm^3, a logging speed of 1,800 ft/h, assuming a 3-level averaging and a 1-σ definition. The vendor also indicates that a simplified relationship between precision and measurement is:

$$\sigma = 0.01 * \rho_{measured}$$

Measurement in the user's well

Two interesting zones are considered, one with $\rho_1 = 2.2$ g/cm^3, the other one with $\rho_2 = 2.75$ g/cm^3. The logging speed has been selected as 3,600 ft/h. The sampling rate is 2 in. The filtering is a 5-level weighted filter with weights 1/10, 1/5, 2/5, 1/5, 1/10. As the user is not comfortable with the 68.3% probability (corresponding to the 1-σ definition), he prefers to obtain a precision value for 2-σ.

Different logging speed and sampling rate

The actual logging speed is twice the reference logging speed. The reference precision is multiplied by √2. The sampling length (2 in.) is 1/3 the reference sampling length (6 in.). The reference precision must be multiplied by √1/[1/3], that is √3.

Different filters

The reference filter corresponds to an improvement of precision:

$$\text{Improvement} = \sqrt{[(1/3)^2 + (1/3)^2 + (1/3)^2]} \text{ or } \sqrt{1/3}\ (0.58)$$

The filter used during logging corresponds to an improvement:

$$\text{Improvement} = \sqrt{[(1/10)^2 + 2*(1/5)^2 + 2*(2/5)^2]} \text{ or } \sqrt{13/50}\ (0.51)$$

Then:

$$\sigma_x = (0.51/0.58) \times \sigma_{xref} = 0.88 \times \sigma_{xref}$$

7. MDT in our example is Measured Density Tool (by contrast to the Formation Density Tool).

Different σ-definition

p = 2 and n = 1. So, the number for precision to be used is twice larger than the reference precision.

Different values (simplified approach)

For $\rho_1 = 2.2\ g/cm^3$, $\sigma = (2.2/2.5) \times \sigma_{ref}$

For $\rho_2 = 2.75\ g/cm^3$, $\sigma = (2.75/2.5) \times \sigma_{ref}$

Final precision values (combining all conditions)

For $\rho_1 = 2.2\ g/cm^3$, $\sigma = (2.2/2.5) \times \sqrt{2} \times \sqrt{3} \times 0.88 \times 2 \times 0.0038 = 0.01442\ g/cm^3$

For $\rho_2 = 2.75\ g/cm^3$, $\sigma = (2.75/2.5) \times \sqrt{2} \times \sqrt{3} \times 0.88 \times 2 \times 0.0038 = 0.01802\ g/cm^3$

Computation with a detailed algorithm

The tool response of the density can be simplified as:

$$\rho = A + B \ln(N_{LS}) - C \ln(N_{SS})$$

where A, B and C are constants depending on the tool. B and C are the sensitivities of the long- and short- spacing detectors. The long- and short spacing counting rates uncertainties are such that:

$$\sigma^2_{NLS} = N_{LS}$$
$$\sigma^2_{NSS} = N_{SS}$$

The partial derivatives of ρ_b versus N_{LS} and N_{SS} are:

$$\partial\rho/\partial N_{LS} = B/N_{LS}$$
$$\partial\rho/\partial N_{SS} = -\ C/N_{SS}$$

Hence:

$$\sigma^2_\rho = N_{LS}(B/N_{LS})^2 + N_{SS}(C/N_{SS})^2$$
$$\sigma_\rho = \sqrt{[B^2/N_{LS} + C^2/N_{SS}]}$$

Numerical application

B = C = 0.5

Reference data set: $N_{LS} = 9{,}800$, $N_{SS} = 14{,}800$,

$\sigma_\rho = 0.0065\ g/cm^3$ without filtering, $\sigma_\rho = 0.0038\ g/cm^3$ with a three-level averaging.

13.4.7 Taking into account the thickness of the zones of interest

If n measurements are available over the same formation, then the precision is improved and represented by a smaller number [4]. In a homogeneous thick formation, these n measurements are collected over n consecutive levels.

$$\sigma_x = \sigma/\sqrt{n}$$

This corresponds to a dramatic improvement of precision (and of the overall uncertainty) in thick beds (Fig. 13.5). The effect of thickness on precision is depicted in Table 13.3.

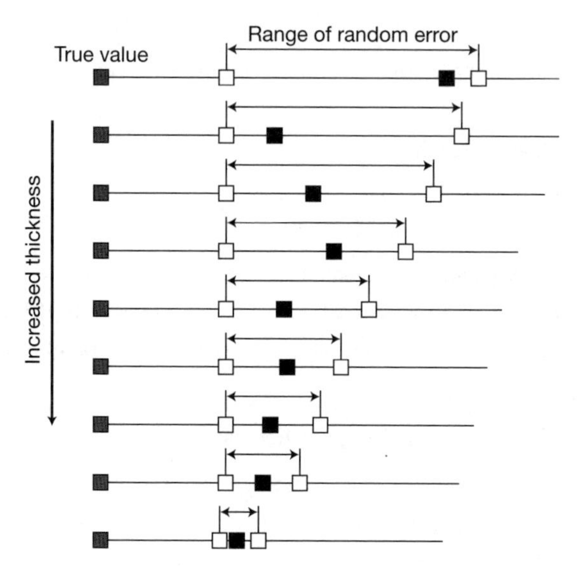

Figure 13.5

Distance from measured value to true value. The black squares between two white squares on the right represent the measured values.

Table 13.3 Improvement of precision with increased bed thickness

Number of levels	Systematic error	Random error	Total error
1	0.01	0.0200	0.0300
2	0.01	0.0141	0.0241
4	0.01	0.0100	0.0200
8	0.01	0.0071	0.0171
10	0.01	0.0063	0.0163
11	0.01	0.0060	0.0160
20	0.01	0.0045	0.0145
50	0.01	0.0028	0.0128

The accuracy specification is defined as 0.01 g/cm^3. The precision specification is 0.02 g/cm^3. All units are in g/cm^3.

13.5 HANDLING SYSTEMATIC ERRORS

As mentioned before, only a few log analysts evaluate the systematic errors. Constant values or vendors specifications are used. In Fig. 13.6, this is represented as the “classical information.” A better way to quantify or at least manage systematic errors is to make full use of all available technical documents, to propagate known errors thanks to correction charts and to monitor the drifts experienced by logging tools.

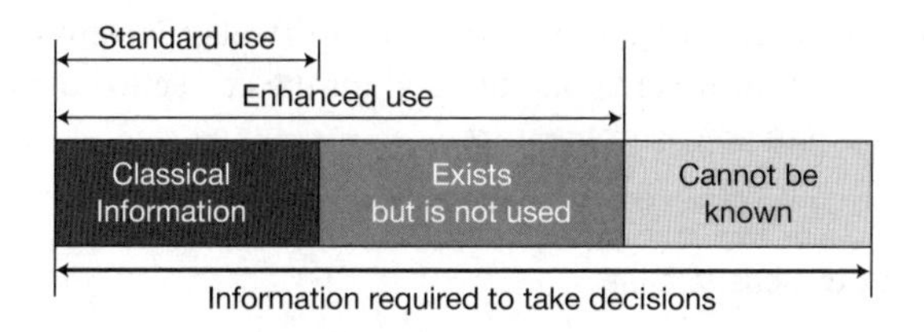

Figure 13.6

Information types.

There is still a limit to completely quantify systematic errors. Information is always limited when compared to the complexity of the logging environment that cannot be described in infinite details. It is also true that a number of job circumstances cannot be perfectly known. What is done about this ignorance: Nothing, but, the experienced data user keeps this limitation in mind.

The next sections describe how systematic errors are better managed through an enhanced use of all available information.

13.5.1 Quantifying propagated errors

Systematic errors, linked to environmental effects, can be compensated if a correction algorithm and an input to the correction are available. It is imperative that all applicable corrections be performed. Even with perfect corrections, using the correct chart and reasonable inputs, an uncertainty balance remains. The inputs to the corrections, whether they come from an external source or from a measurement, are linked to some errors, and hence to some uncertainties. These uncertainties are propagated to the main measurements. To quantify these uncertainties, it is necessary to obtain the sensitivity of the measurement to a given environmental effect. For simple corrections, this is derived from the vendor correction chartbook (Fig. 13.7). The slope of the tangent to the correction curve is the sensitivity.

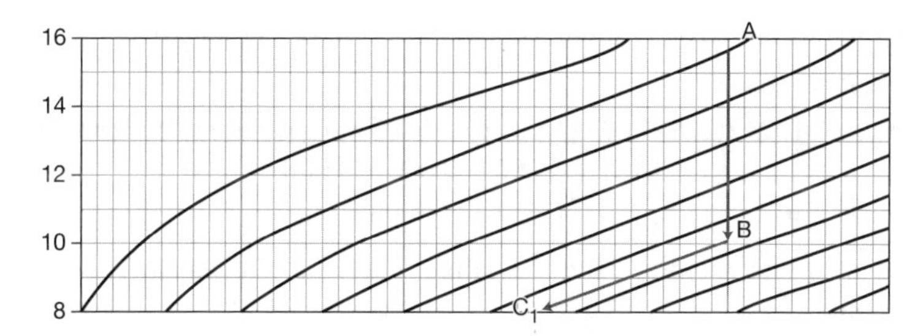

Figure 13.7

Sensitivity to the environmental effect and correction of a neutron measurement to the borehole size. The scale on the x-axis is porosity from 0 to 50 pu. The correction line is straight and the tangent coincides with the correction line. Porosity varies by 11.5 pu when the borehole size varies by 2 in. The sensitivity to the borehole effect is therefore 5.75 pu/in.

A measurement x is affected by p environmental effects. Sensitivities are α, β, γ, etc. The uncertainties on the environmental inputs used for the correction are σ_1, σ_2, σ_3, etc. The propagated uncertainty to the measurement is:

$$\sigma_x^2 = \alpha^2 \sigma_1^2 + \beta^2 \sigma_2^2 + \gamma^2 \sigma_3^2 + \ldots.$$

Example: Neutron porosity measurement

The sensitivities of the neutron porosity measurement to different environmental effects are given in Table 13.4.

Table 13.4 Sensitivity to the environment
Contribution of the different environmental effects

Environmental effect	Sensitivity
Borehole size	1.5/in
Mud cake thickness	3.0/in
Borehole salinity	2.0/(10^6 ppm)
Formation salinity	12.0 (10^6 ppm)
Mud weight (no barite)	0.2/(lbm/gal)
Mud weight (barite)	0.1/(lbm/gal)
Standoff	3.0/in
Pressure	0.2/(kpsi)
Borehole temperature	1.75/(100°F)

The uncertainties on the inputs are obtained from vendors' specifications or from technical articles. For instance:

$$\sigma_{caliper} = 0.25 \text{ in}$$
$$\sigma_{standoff} = 0.10 \text{ in}$$
$$\sigma_{pressure} = 10 \text{ psi}$$
$$\sigma_{temperature} = 5°F$$

The uncertainties on the other inputs are assumed [8] to be negligible.

The propagated error is:

$$\sigma_{\phi N}^2 = (1.5)^2 (0.25)^2 + (3.0)^2 (0.1)^2 + (0.2)^2 (0.01)^2 + (1.75)^2 (0.05)^2$$
$$\sigma_{\phi N}^2 = 0.141 + 0.090 + 0.000 + 0.008 = 0.238$$
$$\sigma_{\phi N} = 0.488 \text{ pu}$$

The propagated uncertainties from the caliper and from the standoff size dominate the overall uncertainty.

8. Assumptions are to be kept to a minimum.

13.5.2 Integrating tool drift in the uncertainty

Vendors' specifications include "reasonable" tool drift. Detectors, photomultipliers and electronic components are particularly sensitive to temperature and vibrations effects, but tool designers excel in minimizing their impact. Drift can be monitored if vendors deliver pre- and post- survey checks, or better, pre- and post- calibrations.

A challenge is to determine where the tool drift occurred. Does it result from a continuous trend, or does it occur instantaneously after a shock or an incident taking place in the hole?

Caliper drift

The caliper drift is quantifiable if the before and after survey checks have been performed by the logging company. When the after survey check is missing, it is possible to use the check in the casing performed at the end of the survey (Table 13.5).

Table 13.5 Example of caliper drift: Caliper 2 shows a shift of 0.4 in.

Remark
Casing weight 133 lb/ft casing ID 18.7 in Caliper check in casing: C1 = 18.7. C2 = 19.1 (offset = -0.4 in)

When the drift exceeds the vendor's specification, it is preferable to use the drift value instead.

Density drift

Electronic drift is generally well controlled in modern density tools. Density drift mostly originates from wear. The design of the tool forces the pad or the sub containing the detectors and source against the formation. This continuous contact causes friction. When the formation is abrasive and/or the survey is long, the pad or sub thickness is reduced, and hence, the tool response is modified.

A note about survey duration in logging-while-drilling

An example immediately underlines the impact of survey duration. A 3,600-ft interval is considered. It takes two hours for a wireline logging tool at a logging speed of 1,800 ft/h to complete a log. It takes 72 h for a logging-while-drilling tool to complete the survey at a rate of penetration of 50 ft/h. LWD tools are therefore more susceptible to suffer from sub wear.

How can wear be detected?

Besides physical observation of wear after a survey, wear can be recognized by comparing successive calibrations (Table 13.6 and Chapter 16).

Table 13.6 Change of counting rates between a pre- and a post-calibration. N-LS-Al represents the counting rates of the long-spacing detector during the calibration in a block (here, Aluminum).

Calibration	Date	Date used	N-LS-Al
			cps
#1	26-july-08	01-august-08	313.40
#2	20-august-08	01-november-03	341.40
Change in %			8.93

Change in density due to calibration drift

A simplified density algorithm is taken:

$$\rho = A - B * \log(N_{LS}/N_{LS\text{–}Al})$$

ρ is the density. N_{LS} represents the long spacing count rate. $N_{LS1\text{–}Al.}$ is the counting rate observed during the calibration in an Aluminum block. B is the long spacing sensitivity. It varies around 0.6. Two densities can be computed, one using the pre-calibration counting rates, the second derived from the post-calibration counting rates.

$$\rho_1 = A - B * \log (N_{LS}/N_{LS\text{-Al-pre}})$$
$$\rho_2 = A - B * \log (N_{LS}/N_{LS\text{-Al-post}})$$
$$\rho_{LS2} - \rho_{LS1} = - B * \log (N_{LS}/N_{LS\text{-Al-post}}) + B * \log(N_{LS}/N_{LS\text{-Al-pre}})$$
$$= B * \log (N_{LS\text{-Al-pre}}/N_{LS1\text{-Al-post}})$$

Taking the numbers from Table 13.6,

$$\rho_{LS2} - \rho_{LS1} = 0.6 * \log (341.4/313.4) = 0.0223 \text{ g/cm}^3$$

It is necessary to allocate the drift at some point in the timing of the survey. Was wear a continuous and smooth process? Or, did it take place suddenly because a given, specially abrasive formation, was traversed by the bit? Short of making assumptions, the most penalizing choice is to consider that the totality of the drift goes into the uncertainty budget.

13.6 ONE STEP FURTHER: ANTICIPATING OTHER SOURCES OF UNCERTAINTY

Once corrections have been performed, correction-induced uncertainties and drifts have been quantified, it is possible to go further. The experienced data user is aware that additional components of uncertainty are present. Some of them can be recognized and quantified.

This approach is developed for the density measurement, but can be performed for any measurement. The experienced data user can develop such a process after having built expertise on logs of the same type.

13.6.1 Density measurement

The density measurement has no correction chart attached to it. So, there is no propagated uncertainty from these charts. Still, it is expected that the following uncertainties do count:

- The uncertainty due to the hole shape and size,
- The uncertainty linked to the mud or mud cake compensation scheme (often called $\Delta\rho$),
- The uncertainty due to the hole rugosity.

The first remark that the log analyst may conclude from the list of uncertainties above is that these uncertainties do not relate to a homogeneous geological zone, but are intimately related to the sensor sensitivity to hole shape and to the mud cake. To quantify the uncertainties, it is necessary to come back to finely sampled data.

These uncertainties are reviewed in sequence [2].

13.6.2 Density uncertainty due to the hole diameter

The hole-related density uncertainty is derived from the slope of the caliper curve.

$$slope_{CALI} = [1/(2*s)]\,[(CALI_n - CALI_{n-1}) + (CALI_{n+1} - CALI_n)]$$

$CALI_n$ is the value of the caliper at level n and s is the sampling rate of the data, expressed in ft. The borehole-size related uncertainty is computed in different ways depending on the caliper and its slope:

If $CALI \leq 9$, $\sigma_{CALI} = 0$

If $9 < CALI \leq 16$, $\sigma_{CALI} = 0.002 * (CALI - 9)$

If $CALI > 16$ and $slope_{CALI} < 0.1$, $\sigma_{CALI} = \sqrt{0.1}$

If $CALI > 16$ and $slope_{CALI} > 0.1$, $\sigma_{CALI} = 0.002 * (CALI - 9)$

13.6.3 Uncertainty originating from $\Delta\rho$

If $\Delta\rho \geq 0$, $\sigma_{\Delta\rho} = 6 * \Delta\rho^2$

If $\Delta\rho < 0$, and $w_{mud} \leq 12$, $\sigma_{\Delta\rho} = 6 * \Delta\rho^2 + |\Delta\rho|$

If $\Delta\rho < 0$, and $w_{mud} > 12$, $\sigma_{\Delta\rho} = 6 * \Delta\rho^2$

w_{mud} is the mud weight expressed in pounds per gallon.

13.6.4 Uncertainty due to the hole rugosity

Uncertainty is quantified by looking at successive caliper values. First a function H_r, hole rugosity, is computed.

$$H_r = (1/4 * s^2) * [(CALI_{n-3} - 2 * CALI_{n-2} + CALI_{n-1}) + (CALI_{n-2} - 2 * CALI_{n-1} + CALI_n) + (CALI_{n-1} - 2 * CALI_n + CALI_{n+1}) + (CALI_n - 2 * CALI_{n+1} + CALI_{n+2}) + (CALI_{n+1} - 2 * CALI_{n+2} + CALI_{n+3})]$$

$$\sigma_{hole\ rugosity} = 0.1 * H_r^2$$

13.6.5 Combining uncertainties

Finally,

$$\sigma_\rho = \sqrt{\sigma^2_{CALI} + \sigma^2_{\Delta\rho} + \sigma^2_{hole\ rugosity}}$$

13.6.6 Uncertainty due to a depth-matching error

The uncertainty due to a depth matching error of 0.5 ft is calculated as follows:

$$\sigma_{depth\ matching} = [1/(4*|s|)]\ [|\rho_{raw\ n} - \rho_{raw\ n-1}| + |\rho_{raw\ n+1} - \rho_{raw\ n}|]$$

An example of an extended computation of the uncertainties is given in Appendix 6.

13.7 VISUALIZATION OF UNCERTAINTIES

Standard log presentation gives a deceptively thin curve. The measured curve, not the real formation curve is displayed (Fig. 13.8). A LAS listing gives even "thinner" values.

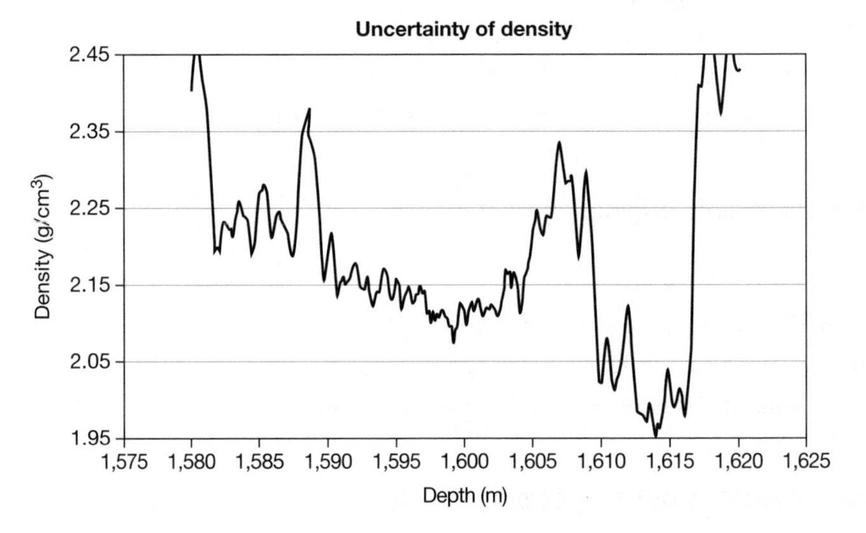

Figure 13.8

Original log information.

The delivered logging curve does not show any uncertainty.

A careful study of uncertainties yields a band of values, not a succession of single values. The band represents the location of the most likely real values of the formation. In the example shown Fig. 13.9, the band does not even include the curve of the measured values.

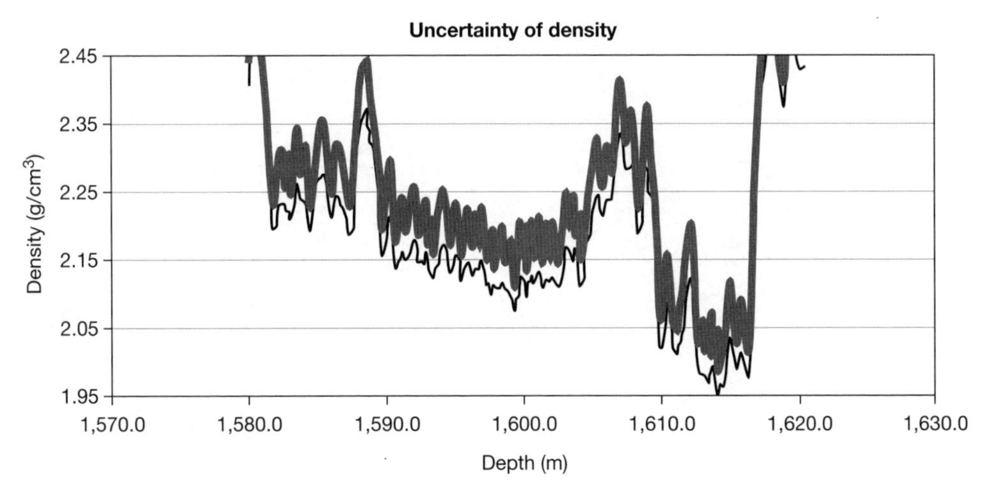

Figure 13.9

Uncertainty band for the same example.

13.8 SUMMARY

- It is possible to manage uncertainties.
- It is a complicated process with many variables and components.
- Vendors' specifications are simplistic. They cannot be used directly.
- Conversion of these specifications to specific job conditions is possible
- The vendors have a large amount of knowledge on the uncertainties of their measurements. This resource needs to be tapped by data users.

REFERENCES

[1] Mayer, C., Sibbit, A., "Global, a new approach to computer-processed log interpretation," paper SPE 9341, 57th annual fall conference and exhibition, Dallas, 1980.

[2] Sibbit, A., Personal communication.

[3] Theys, P., "Log quality control and error analysis, a prerequisite to accurate formation evaluation," 11th European formation evaluation symposium, Oslo, 1988.

[4] Theys, P., *Log Data acquisition and quality control*, Éditions Technip, 1999.

[5] Verga, F., Giaccardo, A., Gonfalini, M., "Determination of uncertainties in water saturation calculations from well log data using a probabilistic approach," 5th offshore mediterranean conference, Ravenna, 2001.

14

Deliverables

Log data? Always too much.
One oil company.

Log data, always too little.
Another oil company.

Get to know the clients better than they know themselves.
Theodore Levitt
(1925-2006)

14.1 IMPORTANCE OF DATA COMPLETENESS

The most difficult challenge in data quality is completeness of the delivery. There is no much point to try to manage uncertainties if the basic data and the auxiliary information to validate it are not available.

14.1.1 Who defines the deliverables?

Intuitively the data users should be the first persons involved in defining logging deliverables. In fact, considering the complexity of the logging environment and the complication of the systems that bring data to the oil company, the onus is more on the vendor side. The company must ensure that all basic and auxiliary data are delivered. The similarity with the purchase of a car has often been used. The buyer does not need to specify to the car manufacturer that a clutch should be part of the car, but the car company definitely needs to make sure that this clutch is designed and present in the car.

14.1.2 The need for graphical displays

The incoherence between the graphical display and the digital data bank, explained in Chapter 8 can be solved by delivering only a complete digital file in the future. Unfortunately, it has been shown that human beings are unable to check long listings of numbers in an efficient manner. Data needs to be organized in well-designed plots and diagrams [9]. Data vendors need to "productize" data, that is, shape it in the most user-friendly presentations for maximum productivity by the user. Productization, a concept well understood in the building of commercial websites, has not arrived yet to the oilfield data arena.

It is hoped that logging companies will be able to deliver complete digital files in the near future and that it will be possible to extract plots and presentations that reflect unambiguously the digital file.

14.1.3 Standardization

The data user faces two fundamental questions:

- Is all that is needed delivered?
- Then, where can it be found?

A strong need for standard deliverables has been expressed, first by the definition of a typical print presentation through API RP 31A [1], [1] then by the definition of a common digital deliverable described in API RP 66 [2].

Unfortunately these efforts have not been continued while the variety of suppliers, measurements and well conditions has exploded. Except for the "main" log, there is no standardization in the presentation of auxiliary information as shown in Fig. 14.1 representing the calibration tails supplied by three vendors. The lack of homogeneity of the displays makes it very difficult to control the quality and validity of the information. As a consequence, only a few data users actually look at the graphical displays. When they do, they only look at the main log. Frequently the logging engineer doesn't know what data to present and, with the rapid changes of the technology, the petrophysicist often doesn't know what data to request.

14.1.4 Format and content

While oil companies feel reassured that they collect data that can be used in the long term because they are using industry **format** standards, they should address the issue of **content**. It is of little use to have the right format if it is not properly populated with complete and structured data. The examples shown in Figs 14.2a and 14.2b indicate that the LAS format, the Log ASCII standard [3], gives no strict guideline on what the files should contain. As a consequence, the logging data records are not complete.

---

1. It is hard to believe that the RP31A standard has been elaborated in 1967 and has not been revised ever since.

Natural Gamma Spectroscopy - D / Equipment Identification

Primary Equipment:		
NGT Cartridge	NGC - C	10
NGT Sonde	NGD - B	8
Auxiliary Equipment:		
NGT Cartridge Housing	NGCH - A	11
NGT Sonde Housing	NGH - C	9
Gamma Source Radioactive	GSR - U	7

Natural Gamma Spectroscopy - D Wellsite Calibration

Background Measurement

Phase	WINDOW 1 Background CPS	Value	Phase	WINDOW 2 Background CPS	Value	Phase	WINDOW 3 Background CPS	Value
Master		130.7	Master		60.87	Master		22.47
Before		84.75	Before		34.62	Before		10.78
After		100.5	After		44.74	After		12.73
	0 (Minimum) 100.0 (Nominal) 400.0 (Maximum)			0 (Minimum) 60.00 (Nominal) 200.0 (Maximum)			0 (Minimum) 10.00 (Nominal) 40.00 (Maximum)	

Phase	WINDOW 4 Background CPS	Value	Phase	WINDOW 5 Background CPS	Value	Phase	SGR Background GAPI	Value
Master		3.447	Master		4.002	Master		40.78
Before		2.251	Before		2.390	Before		24.81
After		3.116	After		3.530	After		30.30
	0 (Minimum) 6.000 (Nominal) 24.00 (Maximum)			0 (Minimum) 10.00 (Nominal) 40.00 (Maximum)			0 (Minimum) 30.00 (Nominal) 120.0 (Maximum)	

Master: May 17 10:21 1992 — Before: May 25 14:16 1992 — After: May 25 16:20 1992

Calibration / Verification Summary

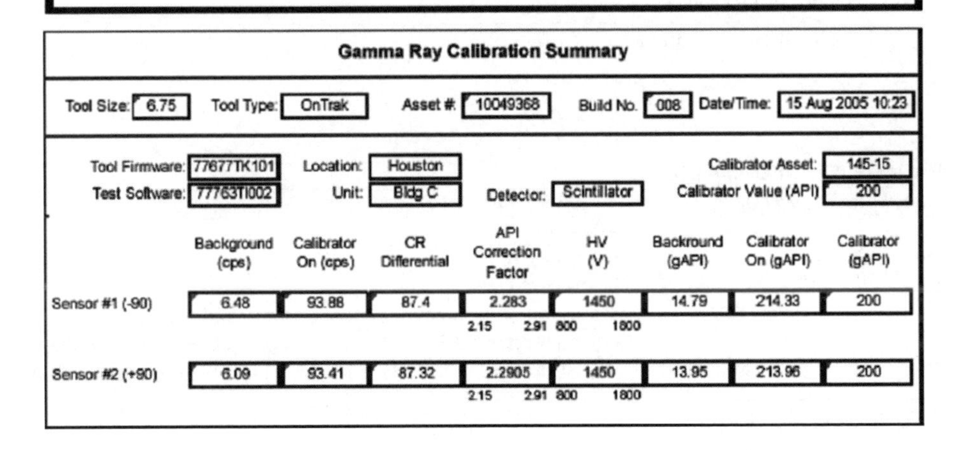

Gamma Ray Calibration Summary

Tool Size: 6.75 — Tool Type: OnTrak — Asset #: 10049368 — Build No. 008 — Date/Time: 15 Aug 2005 10:23

Tool Firmware: 77677TK101 — Location: Houston — Calibrator Asset: 145-15

Test Software: 77763TI002 — Unit: Bldg C — Detector: Scintillator — Calibrator Value (API) 200

	Background (cps)	Calibrator On (cps)	CR Differential	API Correction Factor	HV (V)	Backround (gAPI)	Calibrator On (gAPI)	Calibrator (gAPI)
Sensor #1 (-90)	6.48	93.88	87.4	2.283 (2.15 2.91)	1450 (800 1800)	14.79	214.33	200
Sensor #2 (+90)	6.09	93.41	87.32	2.2905 (2.15 2.91)	1450 (800 1800)	13.95	213.96	200

Figure 14.1

Calibrations – Top: Schlumberger. Bottom: Bakeratlas.

The three records relate to the same information but have nothing in common.

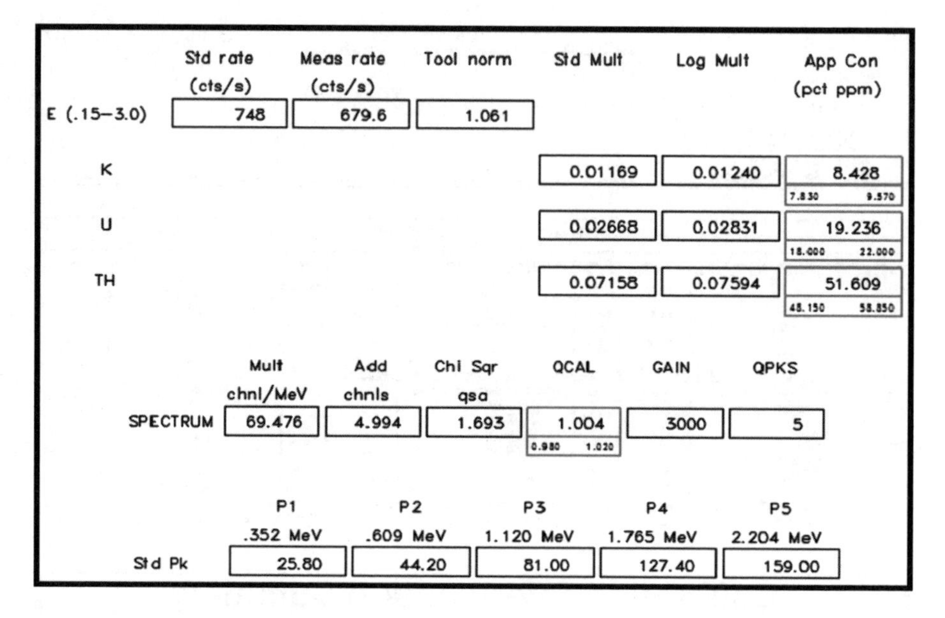

Figure 14.1 (*continued*)

Calibrations – Halliburton.

~PARAMETER INFORMATION		
#MNEM.UNIT	VALUE	DESCRIPTION
#--- ---	------------	------------
RUN .	1	:RUN NUMBER
PDAT .	MSL	:Permanent Datum
EPD .F	0.000000	:Elevation of Permanent Datum above Mean Sea Level
LMF .	ODF	:Logging Measured From (Name of Logging Elevation Reference)
APD .F	684.799988	:Elevation of Depth Reference (LMF) above Permanent Datum
#--		
~CURVE INFORMATION		

Figure 14.2

Examples of poorly filled in LAS file.
There is hardly any parameter information!

VERS	CWLS LOG ASCII STANDARD-VERSION 1.20	
WRAP	NO: One line per depth step WELL INFORMATION BLOCK	
#MNEM. UNIT	DATA TYPE	INFORMATION
STRT.FT	16270.0000	:START DEPTH
STOP.FT	22781.0000	:STOP DEPTH
STEP.FT	0.50000	:STEP LENGTH
NULL	–999.25000	:Null value
COMP.	COMPANY	:APRICOT, INC
WELL.	WELL	:OCS-G-47857#2 ST1
FLD.	FIELD	:SHALIMAR GARDENS 783
LOC.	LOCATION	:
PROV.	COUNTRY	:FLORIDA
NATI.	COUNTRY	:USA
DATE.	LOG DATE	:
API.	API NUMBER	:70-941-30111-02
SECT.	SECTION	:NA
TOWN	TOWNSHIP	:NA
RANG.	RANGE	:NA
DAY.	LOG DAY	:
MONT.	LOG MONTII	:
YEAR.	LOG YEAR	:
PDAT.	PERMANENT DATUM	:
ENG.	LOGGING ENGINEER	:JONES
WIT.	LOGGING WITNESS	:
LATI.	LATITUDE	:
LONG.	LONGITUDE	:
BASE.	LOGGING DIS	:8305

Figure 14.2 *(continued)*

Second example. If not as empty as the previous example, this record does not contain enough auxiliary information.

14.2 CONTENT OF THE GRAPHICAL FILES

When the main log, always present in the delivery package, is submitted to the oil company, the following questions still require some answers:

1) Where were the measurements collected?
2) What was the logging environment?
3) How was it corrected for?
4) Which tools were used?

5) Were they calibrated?
6) Did they function properly?
7) Were the measurements precise?
8) Was any anomalous reading detected?
9) Was the logging data checked?
10) Was any anomalous circumstance of the logging job reported?

These questions are answered through the 14 log components listed in Table 14.1. The number in the second column relates to the recommended position of the component in the graphical presentation.

In addition, some time-based plots may be included. Time-based records are useful for pressure tests and for logging-while drilling operations. They demonstrate that the formation parameters are changing with time. Also optional, but recommended, is the time-depth plot. It establishes the traceability of the transform from time to depth in logging-while-drilling.

Table 14.1 The components of the graphical display answer the questions of the data user.

1. Where were the measurements collected?	(4) Depth Information Box (6) Well Sketch (7) Well Plot (8) Survey Listing
2. What was the logging environment?	(1) Header
3. How was it corrected for?	(9) Parameter Listings (10) Parameter Change
4. Which tools were used?	(5) Tool Sketch
5. Were they calibrated?	(11) Calibration Info Box
6. Did they function properly?	(13) QC Log/QC plots
7. Were the measurements precise?	(12) Repeat Passes
8. Any anomalous reading was detected?	(12) Repeat Passes
9. Was the logging data checked?	(14) LQC box
10. Was any circumstance of the logging job reported?	(2) Remarks Section (3) Log Chronology

14.2.1 Depth-related information

Depth is often taken as granted. In fact, the related information is complicated and is reviewed in details in Chapter 15.

14.2.2 Tool sketch

The tool sketch gives a visual display of the combination of logging tools. Auxiliary equipment, such as knuckle joints, standoffs and excentralizer is shown as its presence is essential to the acquisition of accurate information. The tool serial numbers are important to verify the match with calibration data. It is also useful to have these numbers to track rogue tools giving erroneous data in a consistent manner.

14.2.3 Remarks

Remarks are important and must be used for nonstandard circumstances dealing with tool operation, log presentation, anomalies, special requests, customer orders or authorizations that change the standard operational procedures. Low-content remarks, such as:

- Tool run as per tool sketch.
- Logging company procedures are applied.

Need not to use the precious real estate.

In particular, it should be clearly indicated when tool response or environmental corrections are no longer valid or when the logging tool has reached the limits of its operating specifications. This is the case with the induction borehole correction that is not defined when the hole diameter is above 22 in. It is also important to highlight the fact that volume integrations are not accurate when caliper measurements reach their limit (caliper fully extended, not touching the formation).

Figure 14.3 displays an example of well-reported remark. The logging engineer describes how wear affects the logging measurement in a dramatic manner. The data user definitely needs to be made aware of these corrections. The corresponding digital file does not contain this critical information and cannot be used for quantitative information.

Data density on the log was low due to fast drilling over long intervals. Memory log was of good quality.
Due to wear on the density sub, the sub was post-calibrated. The density with the new calibration was 0.043 g/cm^3 higher than with the old calibration. The wear has been adjusted as a function of the density correction which is observed to be increasing during the run.
The density correction starts at 0.008 and increases to 0.047 g/cm^3 at 3,835 m. At 3,835 m it has an abrupt change to 0.065 g/cm^3.
From 3,835 m to 4,165 (last reading) it increases to 0.085 g/cm^3. Half of the wear is put on the section down to 3,835 m. This means that 0.022 g/cm^3 had to be accumulated in this interval.
0.01g/cm^3 was put as an abrupt shift at 3,835 m and the rest (0.022 g/cm^3) from 3,835 m down to total depth.
The following formula is used to correct the density:
First interval: New density = old density + 0.0225*[(x – 609)/(3,835 – 609)]
where 609 m is the depth of the first density point of the section.
Last interval: New density = old density + 0.0125*[(x – 3,835)/(4,174 – 3,835)] + 0.0325
where 4,174 is the depth of the last density point in the hole section and 0.0325 g/cm^3 is the amount of wear that had affected the density data down to 3,835 m.
Due to the wrong density rib in the offshore software, the original density is not exactly the same as the...

Figure 14.3

Example of extensive remarks.

14.2.4 Job chronology

Every logging job is different [2]. The circumstances need to be reported by the logging engineer in a chronology of the job. The valuable information helps considerably the data user. It is recommended that the field engineer uses a wording comprehensible by the non-expert.

Monday, November 11th, 2010	
01:30	Disconnect ABT from CDT string to check it alone
02:30	Start checking ABT
03:00	Start ABS 49 base oil station logs
03:30	Start ABS 48 OBM station logs
04:00	Do simulation logs with parameters as per logging program
05:00	Start checking ABS 173
06:00	Finish checking ABS 173 ---> USE THIS AS MAIN
06:10	Start checking Packer tool
06:45	Finish checking Packer tool
10:30	Finished calibrating tension device TD 1234 for 7-48 and 7-
11:30	46 cables
14:00	Checked swivel head 567 with mega-ohm-meter
15:00	Operations and safety meeting with contractors personnel
16:20	onboard
16:55	EFX125-GHT string ready for operational check
17:20	EFX125-GHT OK
17:25	Communication with onshore checked. OK.
20:00	Netsighter checked OK.
21:00	Announcement made to stop all hot work due to gas in mud
21:05	above 184
21:30	Start checking backup EFX
22:00	CANNOT CLOSE EFX caliper. Start troubleshooting.
22:15	Restart surface system; Still cannot close EFX caliper.
22:20	Check continuity and insulation on logging head. Checks are
	OK, but EFX still fails.
	Cable trim. EFX still fails
	BTMS 3660/BTMD 3683/BTCC 3675 FAILED
	Start checking backup CDT modules.
	......

Figure 14.4

Example of a full description of the logging job (this is a partial extract).

2. The word "log" is derived from the Navy. The logbook, specific to every ship, collected all the events and adventures lived by the crew. The well log depicts what the tool is seeing in the hole. It also includes some part of adventure.

14.2.5 Parameter summary and parameter change

The production of logging curves requires extensive processing controlled by a large number of options and parameters. These parameters are listed in long tables. They contain enigmatic acronyms. Fig. 14.10 shows one of the simplest documents, the list for a Schlumberger Array Induction Tool. Some comments, in the frame, have been added. Parameters fall in three main categories:

- The ones that the data user feels as very important (tool standoff, temperature, etc.),
- The ones left to default value (e.g., – 50,000 m). Hopefully, these parameters have no impact on the processing.
- Some less important parameters.

Some parameters require expert knowledge and are very important:

- AIGS: Akima Interpolation Gating.
- ABHM: Induction Borchole Correction Mode. It can be the most important of them all.

Therefore, they are of little use to the petrophysicist, without an immediate access to a "help" file. The logging company is required to communicate the list of parameters, but it has to be delivered in a user-friendly manner.

These parameters should be displayed in the form of a block diagram that describes their impact on the data. Figure 14.6 is an example of a possible presentation.

An uninterrupted data flow, from raw data to usable petrophysical parameters would enable the data user to reconstruct the processing of data. There are two major obstacles to this. First, a lot of processing is proprietary and black boxes can be found over the data flow. This means that only the data acquisition company controls the flow charts describing what is happening to the data. Second, it is not guaranteed that the processing parameters have been captured and transmitted to the database. This is particularly true for interpretation decisions that are rarely described.

14.2.6 Raw and QC curves

There are many curves (channels) that represent Quality Control (QC) on the raw data. They are essential should the desire to reprocess the data surface in a future time would arise. The first quality curve introduced in the industry is the density correction curve. It became so popular that its absence from the log was quickly noticed by the log analyst. The QSS and QLS curves of the Litho-Density tool were widely used in the 1990s.

When a new tool is designed, if often happens that not all the environmental corrections are available. This is the case of eccentricity and apparent dip effects on the induction tools. These limitations can be overcome by re-processing, but only if raw curves are available. Those curves are most often not requested or provided to the customer, possibly not even recorded by the logging company.

The overwhelming recommendation is that all raw curves and all quality control curves, linked to the documentation allowing their use, [6] and [7], should be delivered to the oil company. This cannot be over-emphasized. It must become mandatory and standard.

Name	Description	Value	Comment
ABHM	Array Induction Borehole Correction Mode	0_ComputeMudResistivity	Very important
ABHV	Array Induction Borehole Correction Mode Version Number	900	???
ABLM	Array Induction Basic Logs Mode	6_One_Two_and_Four	???
ABLV	Array Induction Basic Logs Code Version Number	223	???
ACDE	Array Induction Casing Detection Enable	No	OK
ACEN	Array Induction Tool Centering Flag (in Borehole)	Eccentered	Very important
ACSED	Array Induction Casing Shoe Estimated Depth	-50,000 M	Left to default value
AETP	Array Induction Enable Sonde Error Temp&Press Corr		Very important
AFRSV	Array Induction Response Set Version for Four ft Resolution	41.70.24.20	???
AIGS	Array Induction Select Akima Interpolation Gating	On	Important
AMRF	Array Induction Mud Resistivity Factor	1	Left to default value
AORSV	Array Induction Response Set Version for One ft Resolution	41.70.24.20	???
ARFV	Array Induction Radial Profiling Code Version Number	701	???
ARPV	Array Induction Radial Parametrization Code Version Number	232	???
ARTS	AIT Rt Selection (for ALLRES computation)	AITM_TwoResTrueDeep	???
ASTA	Array Induction Tool Standoff	1.5 IN	Very important
ATRSV	Array Induction Response Set Version for Two ft Resolution	41.70.24.20	???
ATSE	Array Induction Temperature Selection (Sonde Error Correction)	Internal	Very important
AULV	Array Induction User Level Control	Normal	???
AZRSV	Array Induction Response Set Version for Z Resolution	00.10.25.00	???
BHT	Bottom Hole Temperature (used in calculations)	112 DEG	Very important
FEXP	Form Factor Exponent	2	Left to default value
FNUM	Form Factor Numerator	1	Left to default value
GCSE	Generalizes Caliper Selection	HCAL	Very important
GDEV	Average Angular Deviation of Borehole from Normal	19 DEG	Is it used?
GGRD	Generalized Temperateure Selection	0.018227	Important
GRSE	Generalized Mud Resistivity Selection	CHART_GEN_9	Important
GTSE	Generalized Mud Resistivity Selection	HRTS_HTEM	???
RTCO	RTCO – Rt Invasion Correction	YES	Important
SHT	Surface Hole Temperature	20 DEGC	Important
SPNV	SP Next Value	0	???

Figure 14.5

List of parameters for the Schlumberger Array Induction Tool. Comments have been added in the last column.

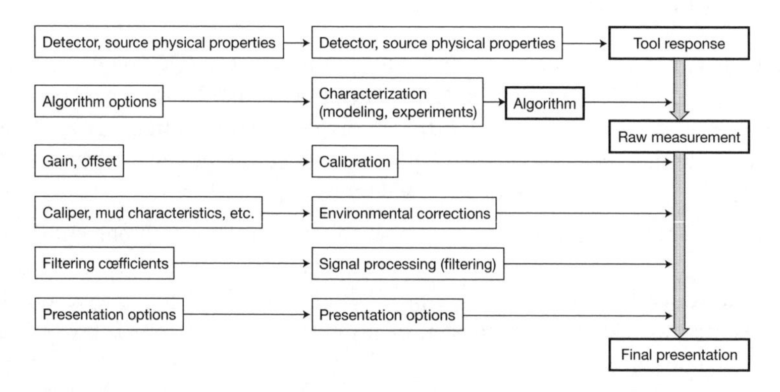

Figure 14.6

Example of simplified block diagram – An indispensable component in the data delivery.

14.2.7 LQC stamp

Most organizations manufacturing complicated products include a quality stamp in their delivery. The stamp indicates that the product has been checked in a systematic manner. Most of the checks are common to all logs, whether they are acquired in open-hole or cased-hole, by wireline or while-drilling conveyance. They include a review of:

- Calibrations
- Corrections
- Effects of environment
- Depth
- Proper auxiliary equipment
- Availability of all supporting information
- Specifications

Additional checks, specific to the measurement, are required. These checks are often arcane. Information on them is available from the logging company.

14.3 DIGITAL FILES

14.3.1 The basic digital file

A basic principle applies to digital files. Anything shown on graphical files should be included in the corresponding digital file. All the items listed in the previous section should be echoed on the digital file. The following paragraphs concern additional components.

14.3.2 Raw data on digital files

All curves for every job should be recorded so that it is possible to retrieve and utilize the data quickly and easily. There are many obstacles to this that exist: ownership, storage and retrieval costs, national, state, and local regulations. [3]

14.3.3 Time based data

As seen in Chapter 8, the logging-while-drilling tools observe the underground formations several times on successive runs. Recent analyses [5] prove that, even on a single given drilling run, the drill string is advancing only 30% of the time in the best case (Fig. 14.7). As a consequence the LWD sensors are scanning the formation over and over. In some extreme

3. In the Petroleum Industry, this need for rapidly accessible data also goes well beyond logs.

cases, when the borehole needs to be worked on, and the bit pulled up and down, the percentage is lower (Fig. 14.8).

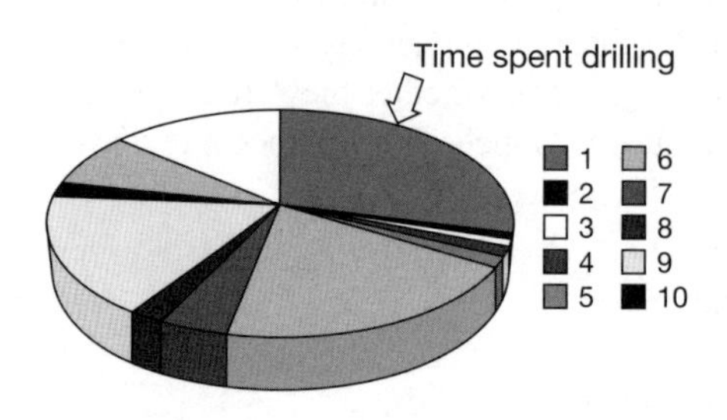

Figure 14.7

Data extracted from the analysis of drilling tasks.
By courtesy of SPWLA.

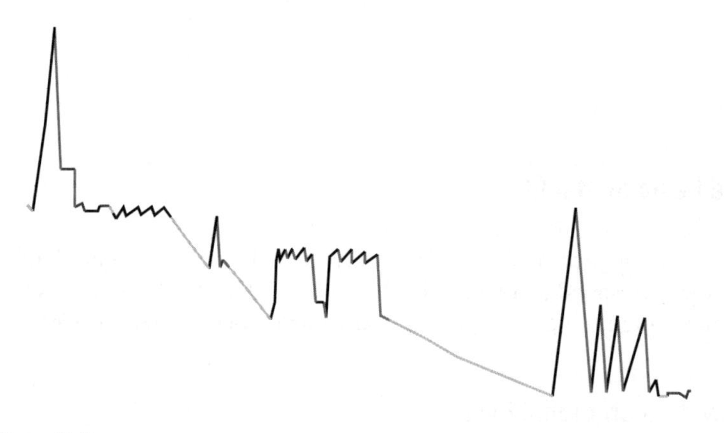

Figure 14.8

Depth-time diagram.
Depth is on the y-axis, time on the x-axis. Only the grey line corresponds to a progress in drilled footage. The driller tries to keep the hole open.
By courtesy of SPWLA.

In these conditions, there can be many different formation measurements at the same depth. A depth-sampled file, allowing only one measure per depth sample is inadequate. Time files become necessary. They can bring interesting insight in the behavior of the formation as shown in Fig. 14.9.

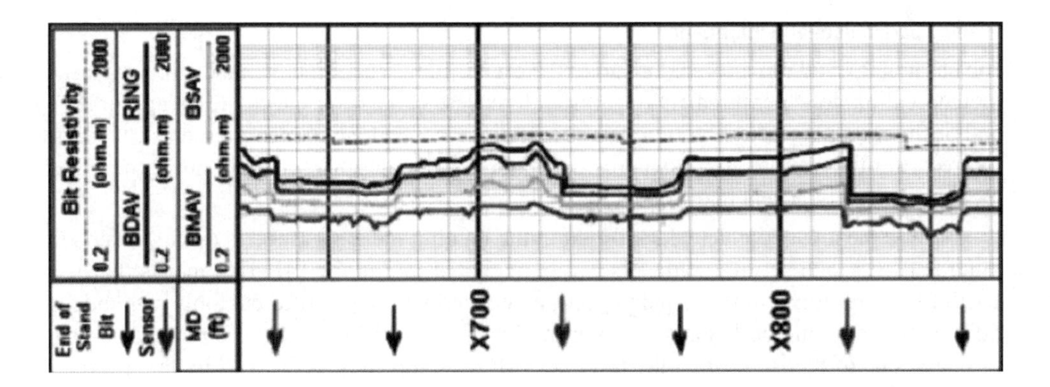

Figure 14.9

Example of four resistivity passes showing the opening of a fracture.
By courtesy of SPWLA.

14.3.4 Data in digital files

Digital files may now contain terabytes of data. But, how can it be used? All deliverables should be standardized, which means that, once the data has been acquired, the oil company package should not require any human intervention. All data components should be delivered regardless of the acquisition company or the oil company. Transparency in the data flow is essential. The perusal of log data in its best form, DLIS, is of extreme complication. It cannot be checked and verified easily. The interrelation between data objects is often not obvious. Software tools should be developed to show how different components of information relate to each other. Data flow diagrams as explained in the previous section should be available along with the data. Similar tools should also be available in databases. For instance, a density curve needs to be always linked to borehole conditions as it is strongly affected by them.

Finally, one should refrain from using any data unless it is weighted with some kind of quality or/and uncertainty rating. Any useable data channel should have for companion a quality flag, or better, an uncertainty bar.

Missing files

The final digital delivery includes only a few of the job acquisition files. These "missing" files are seldom archived by the logging company and are therefore lost forever. Those files include the repeat sections and the depth correlation records.

Missing parameters and information

Job chronology and remarks seldom go to the digital file. In addition, while calibration tasks include calibration measurements against references, calibration checks (plateau checks), and/or the recording of calibration data (coefficient, context, shop measurements), the digital record often capture empty records. [4] Parameter Changes are generally not recorded as most formats only accept one value per parameter.

DLIS is the most complete data format in the logging industry. But it is useful to keep in mind that only one set of parameters can be included in one file [4]. This means that if a DLIS file corresponds to five separate runs, associated with three different tools with indeed different calibration coefficients, varying mud properties and changes of bit sizes for each run, only one of the runs, often the last one, will be completely defined while the information describing the four other ones will be omitted.

14.4 THE IMPORTANCE OF CONTEXTUAL INFORMATION

If contextual information is not recorded and made available to the data user, log data becomes rapidly useless, especially years after the time of acquisition when it is not possible to get in touch with the logging engineer [8]. Through the processing of log interpreters, data may sometimes tell the opposite of what it is supposed to say.

Contextual information is all the data around the main data, the later one generally displayed versus depth. Computer technologists call contextual information metadata. The porosity log run in a deviated well is taken as an example. At 12,456 ft, measured depth, the value is 23.45 pu. A few questions need to be asked:

- Is this a density-, neutron-, NMR-, acoustic-derived porosity?
- Is it corrected? If yes, what are the corrections?
- Is it calibrated? If yes, what are the calibration coefficients?
- Was the tool properly centralized? With what? What is the size of the centralizing device?
- Is there some signal processing? What are the processing options?
- Where is the volume investigated by the logging tool located?
- Is depth documented? Is the depth-measuring device calibrated? Is depth stretch-corrected? Is there a stable procedure to acquire depth?
- How was deviation and azimuth measured? If this is through a magnetic survey, how was magnetic declination corrected for? What is the geodetic reference?
- Which logging company ran the log? Who ran the log? What was the level of training of the field engineer?

4. Thoroughly checking a DLIS file requires a rare expertise. It can be found that many data boxes are actually empty.

The informed user can reach a better understanding of data through the perusal of contextual information. This information allows a buildup of confidence in the data and a subsequent reduction of the risks.

14.5 SUMMARY

- Much attention is given to format, but not enough to **content**.
- Most logging records are incomplete.
- There is no standardization of log deliverables.
- Raw data and QC curves are to be imperatively delivered to oil companies.
- The multiple and different products are extremely difficult to be checked.

REFERENCES

[1] API RP 31A, *Standard Form for Hardcopy Presentation of Downhole Well Log Data*, 1967.

[2] API RP 66, *Recommended Practices for Exploration and Production Data Digital Interchange*, 1996.

[3] Canadian Well Logging Society, *Floppy Disk Committee, Log ASCII standard*, 2009.

[4] Citerne, A., Personal communication.

[5] Maeso, C., “Invasion in the time domain from LWD resistivity: an untapped wealth of information,” SPWLA 51st annual logging symposium, Perth, 2010.

[6] Schlumberger-Anadrill, *Schlumberger, Log Quality Control Reference Manual, MWD & LWD,* 1994.

[7] Schlumberger, *Log Quality Control Reference Manual, Wireline*, 1992.

[8] Theys, P., “Keeping things in context,” *Hart’s E&P,* vol. 79, n° 8, pp. 63-64, 2006.

[9] Tufte, E., *The visual display of quantitative information,* Graphics press, Cheshire, 1983.

[10] Wikipedia, Image file formats: en.wikipedia.org/wiki/Image_file_formats.

15

Depth

15.1 IMPORTANCE OF DEPTH

Depth is the most important log measurement. It impacts oil companies in many different ways:

- Accuracy of absolute depths is required to test the hydraulic continuity between different wells and to verify the lateral continuity of geological formations. It is an indispensable input to the understanding of the geometry of the reservoir.
- Reservoir thickness is derived from depth measurements. It directly impacts the hydrocarbon volume in place. An error of x% in thickness is directly reflected as an error of x% in hydrocarbon volume in place.
- Depth correlation is used to accurately perforate the producing zones.

Depth is associated with directional surveys (including deviation and azimuthal data) to provide the data user with a full description of the location of the rocks and of the reservoir. [1]

15.1.1 Examples of challenges proposed by depth measurements

There are many internal case studies that have been performed about depth [10]. Most are unfortunately not public. A few articles contain the analysis of depth differences between wireline logging runs [8]. Values of 24 ft at 9,000 ft are mentioned. Even the thickness, a "relative" depth as it considers the top and the bottom of the reservoir, is difficult to rigorously evaluate. A specific example shows a difference of 22 ft on a 130-ft thick reservoir surveyed with "wireline" and "logging-while-drilling" conveyed logs. This represents an error of 17%, directly impacting the volume of hydrocarbons in place.

Operating practices related to depth have changed much in the last two decades. While the true depth of a geological event does not change, the measured depth is different every

1. For more information on directional surveys, consult references [11], [12] and [13].

time the processes introduced by the logging companies are modified. These operating changes create geological artifacts that are not actually present.

In the example referenced in [9], the difference between driller's depth and logger's depth in Tuscaloosa logging jobs run from 1977 to 1999 was less than 10 ft (at depths in excess of 21,000 ft). Since 1999, the difference between logger's depth and driller's depth ranged from 22 ft at 15,000 ft to 48 ft at 23,000 ft.

In old oil fields, there is a definitive resistance to modify the historical geological model, even though data confirms that recent depth measurements are more accurate. First, there is often insufficient evidence to corroborate the correctness of the log depth because there is a lack of audit trail. Second, the petroleum engineer or operation geologist is faced with a difficult choice to assume large errors in the log depths to fit the logs to the reservoir model or to drastically change the reservoir model to conform to the logs.

15.2 THE DIFFERENT DEPTHS

The data user has to understand that different measured depths are available from data vendors – and from the driller. It is also important to keep in mind that none of these depth measurements is identical to true depth.

15.2.1 Wireline mark-derived depth

Magnetic-mark-derived depth has been available since the 1950s on logs using wireline conveyance. Magnetic marks are created on the cable at regular intervals (every 100 ft or 50 m). This is done on surface, under a fixed tension (generally 1,000 lbf), with a "ruler" of fixed length (e.g., 12.5 ft). Marks are kept constant during logging. For instance, the first mark is detected close to surface at 83.2 ft. Then every time a magnetic mark is detected, the odometer is adjusted [2] to bring the reading back to 83.2. At total depth, a correction for stretch may be added to change the mark as a block shift.

There are a number of limitations to this type of depth. Marking and mark detection are performed at surface. Downhole cable behavior is not accounted for. Temperature effects, at marking stage and during logging, are also not accounted for. There is inelastic stretch or shortening between marking and logging. Cable is marked at 1,000 lbf while, during logging, it is under a different tension. "Cranking" is not traced.

2. In logging jargon, the depth is said to be "cranked."

15.2.2 Wireline calibrated-wheel depth

In this second method, depth is directly measured by precision wheels located at surface (Fig. 15.1).

Figure 15.1

Schlumberger dual wheel device.
By courtesy of Schlumberger.

These precision wheels are connected to encoders. The wheels are calibrated every six months with a calibrator (Fig. 15.2).

A correction for stretch is quantified with a documented procedure. The tool string is carefully lowered to total depth (TD) at normal speed. A short section, generally 200-ft long, close to TD, is logged down. The section includes distinctive formation characteristics, for future correlation. At TD, the calipers are opened and a section overlapping the down log is logged up. A difference between the sections logged up and down is observed. It is defined as an adjustment. The depth system is corrected accordingly. The adjustment length is archived for traceability. The whole logging interval is then surveyed up without cranking.

Possible errors include effects of wear and slippage of the wheels. Wear is compensated and controlled through a calibration that carefully measures the diameter of the wheels. Slippage is controlled by a dedicated processing, called "fast-wheel algorithm." It is used when the two wheels are not reasonably agreeing. In icy conditions, it may happen that the wheels do not turn while the cable moves by tens of ft [5]. This is hopefully observed by the field engineer who needs to report this condition. There may be some incomplete corrections for elastic stretch (log "down" and log "up"). Pressure, temperature and yoyo effects are not fully accounted for. Small displacements of the tools run for images may be compensated by

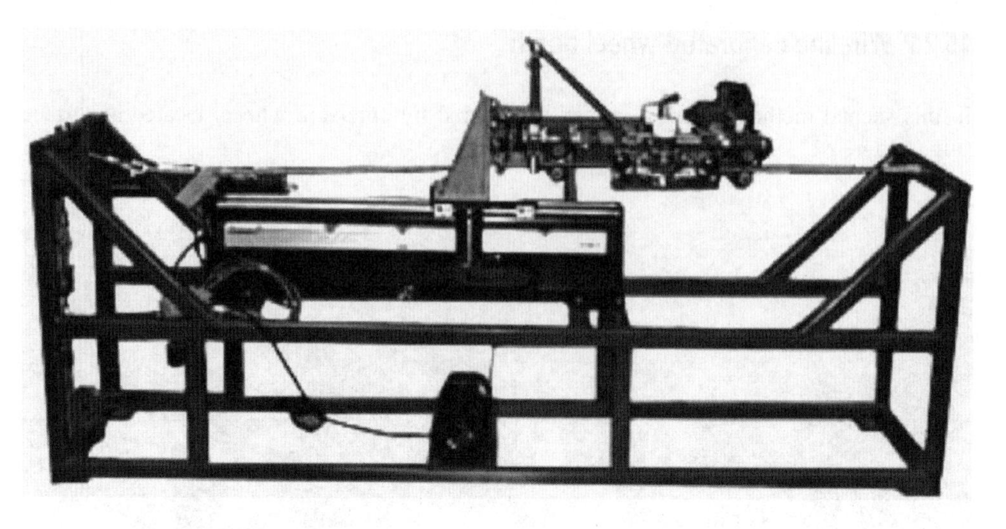

Figure 15.2

Depth wheel calibrator.
By courtesy of Schlumberger.

speed correction processing. The "wheel" methodology is not perfect but has the great advantage of being transparent.

15.2.3 Driller's and LWD depth

Drilling crews also collect information about depth. They measure the length of each drill string component. This is achieved with the pipes being stacked vertically or laying horizontally on racks. While drilling, the depth is computed as the sum of all these individual measurements. By convention, driller's **depth is not corrected for anything** while mechanical and temperature stretches, ballooning and buoyancy effects have been recognized and sometimes quantified, [1] and [3].

Logging-while-drilling companies do not measure depth directly. They align depth to driller's depth at every connection performed on the drill string. Indeed, they have a process to produce a continuous depth between these connections, so that a log display can be supplied versus depth. As there are small differences between the LWD depth system and driller's depth, shifts are performed at the drill string connection. These shifts need to be carefully documented as shown in Fig. 15.3.

Stand number	Joint length	Driller's depth	LWD depth	Difference	Adjustment
s	32.25	9,465.96			
d	32.54	9,498.50	9,499.40	0.90	– 1.00
67	32.43	9,530.93	9,533.20	2.27	– 2.50
s	30.66	9,561.59	9,562.13	0.54	
d	31.78	9,593.37	9,594.90	1.53	– 1.50
68	32.46	9,625.83	9,625.85	0.02	
s	30.62	9,656.45	9,659.00	2.55	– 2.50

Figure 15.3

Listing of the LWD depth shifts. The unit is ft.

Adjustments performed by the LWD engineer are not strictly equal to the depth difference between driller's depth and LWD depth. In some cases, no adjustment is made at all. Adjustments may directly impact the thickness of the reservoir.

15.2.4 Expected differences between the measured depths

Mark-derived- and precision-wheel- depths are different. There has been little experimentation to quantify the differences, even though it should be a simple endeavor consisting in performing the log while enabling the two measurements.

Wireline depths (of the two types) are corrected for stretch, but driller/LWD depths are not. At 10,000 ft, the wireline depth is longer (or deeper), by about 12 to 15 ft. The difference is not random and can be quantified, once the hole trajectory is known. Two examples are given in Fig. 15.4.

For a straight hole drilled to 10,000 ft, the LWD depth reads 9,986 ft and the wireline depth 10,003 ft. For a deviated well with an absolute depth of 10,000 ft, LWD depth is 9,992 ft and wireline depth 10,017 ft.

Unfortunately, because of the poor understanding on how the depths are derived, some operating mistakes are observed. The wireline depth is adjusted to match the casing shoe depth, or the total depth logger is adjusted to match total depth driller. When a well is surveyed with logging-while-drilling on a shallow run, it is not rare that the subsequent wireline log is depth-matched to the LWD log. As one depth is not stretch-corrected and the other one is, it creates depth errors for all deeper runs.

15.3 IMPORTANCE OF GETTING IT RIGHT AT SURFACE

Acquiring accurate depth starts before the well is begun. Accurate well location and the correct elevation are prerequisites.

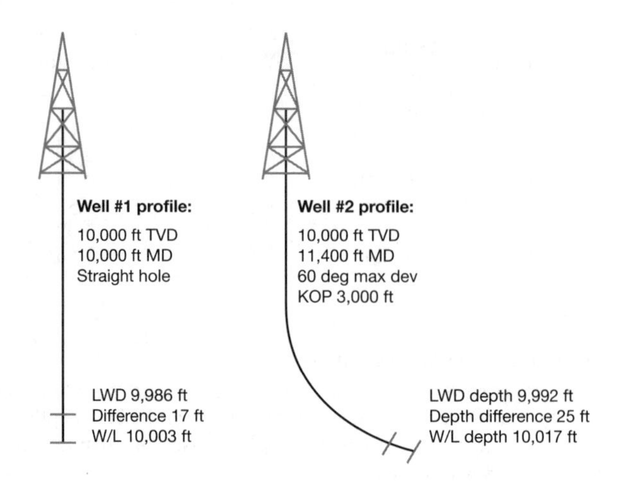

Figure 15.4

Comparison of wireline and LWD depths in two configurations. KOP is the kick off point. W/L is an abbrevation of wireline.
By courtesy of Schlumberger.

15.3.1 Datums

The geographical coordinate system is seldom documented on a log. If this system is not identified, then longitude and latitude may be ambiguous [14]. See Fig. 15.5 for an example.

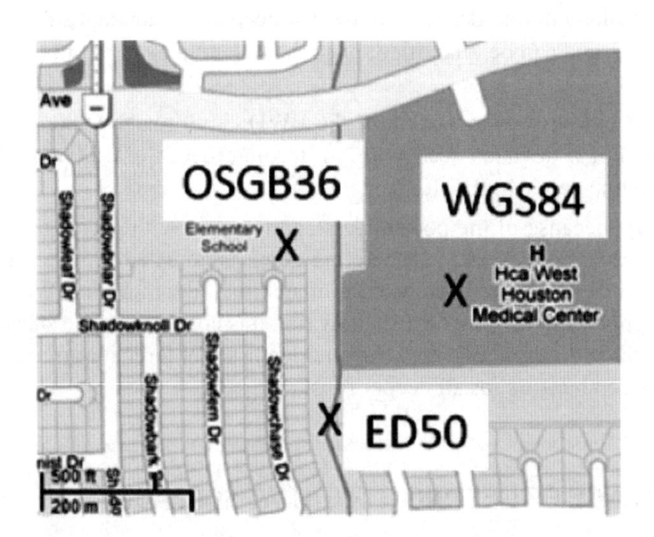

Figure 15.5

Importance of the geographical coordinate system. The three points use three different coordinate systems, called OSG36, WGS84 and ED50.

The three points on the diagram have the same latitude and longitude, but correspond to three different coordinate systems. Imagine that one is the location of an elementary school, the second a hospital and the third one the location of the terrorists' headquarters. The ambiguity could be catastrophic if bombers' pilots are only given the latitude and the longitude, with no specific detail on the geographical coordinate system. A number of wells have been drilled at the wrong place because of similar confusion.

15.3.2 Clear definition of the North reference

North is an ambiguous reference. Lease location is often referred to grid north. Directional surveys are often based on earth magnetism, and are measured from magnetic North. These two "Norths" are different from True North. A clear relationship between the three, as shown in Fig. 15.6 is critical to avoid human errors.

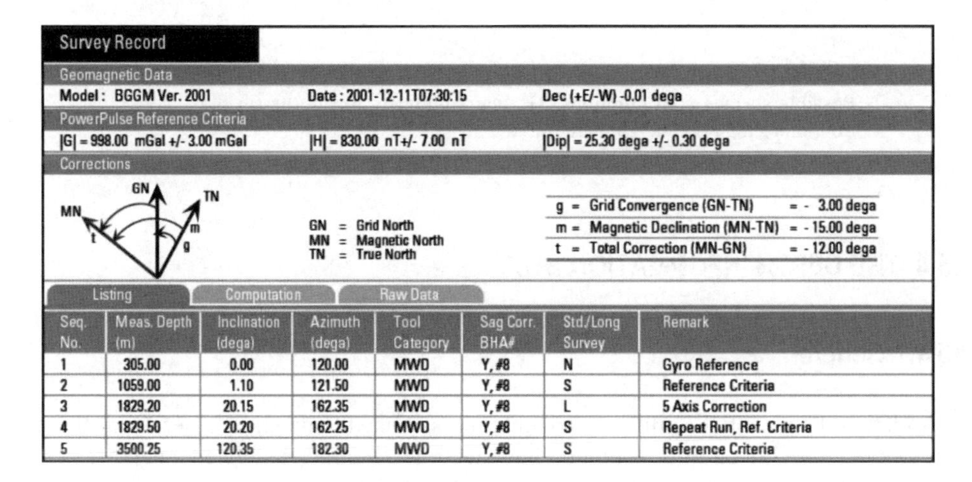

Survey Record

Geomagnetic Data

Model: BGGM Ver. 2001 — Date: 2001-12-11T07:30:15 — Dec (+E/-W) -0.01 dega

PowerPulse Reference Criteria

|G| = 998.00 mGal +/- 3.00 mGal — |H| = 830.00 nT+/- 7.00 nT — |Dip| = 25.30 dega +/- 0.30 dega

Corrections

GN = Grid North
MN = Magnetic North
TN = True North

g = Grid Convergence (GN-TN) = - 3.00 dega
m = Magnetic Declination (MN-TN) = - 15.00 dega
t = Total Correction (MN-GN) = - 12.00 dega

Listing — Computation — Raw Data

Seq. No.	Meas. Depth (m)	Inclination (dega)	Azimuth (dega)	Tool Category	Sag Corr. BHA#	Std./Long Survey	Remark
1	305.00	0.00	120.00	MWD	Y, #8	N	Gyro Reference
2	1059.00	1.10	121.50	MWD	Y, #8	S	Reference Criteria
3	1829.20	20.15	162.35	MWD	Y, #8	L	5 Axis Correction
4	1829.50	20.20	162.25	MWD	Y, #8	S	Repeat Run, Ref. Criteria
5	3500.25	120.35	182.30	MWD	Y, #8	S	Reference Criteria

Figure 15.6

Auxiliary information of the directional survey is shown. In particular, the three "Norths" are clearly represented while an algebraic representation [±] is error prone.

15.3.3 Elevation model

Equally important is the description of the rig position as related to a more durable reference. As shown in Fig. 15.7, more than one number is required.

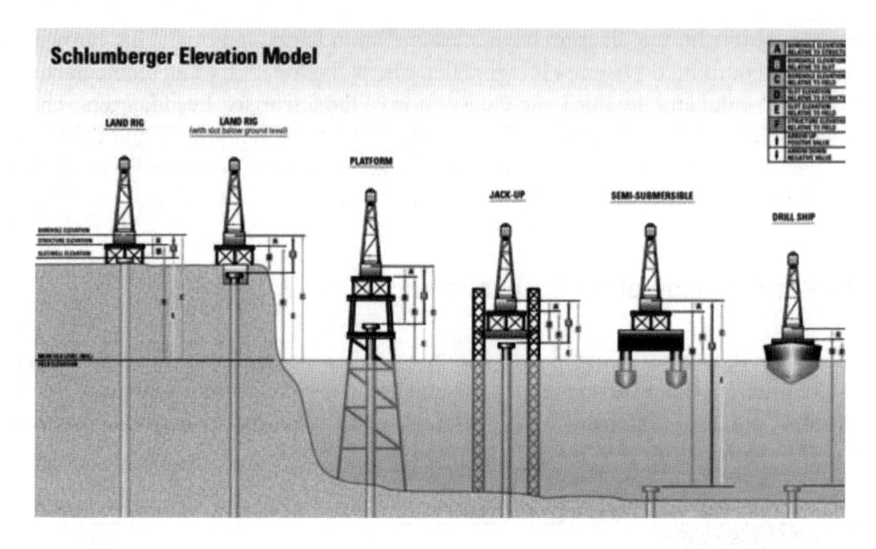

Figure 15.7

Possible surface configurations. To unambiguously locate the rig floor elevation, more than one height is required.

By courtesy of Schlumberger.

15.4 THE DEPTH INFORMATION BOX

15.4.1 General

Considering the potential confusion and ambiguities linked to depth, it is necessary that the data vendors provide all information on how depth is acquired, calibrated and corrected. It can be gathered in "the depth information box," depicted in Fig. 15.8.

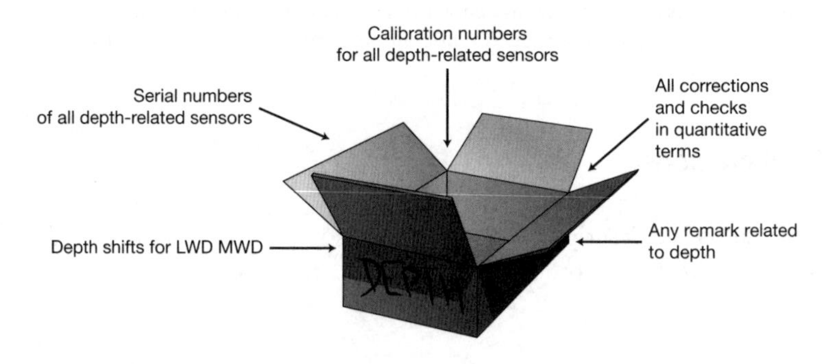

Figure 15.8

The depth box should be part of the log deliverables.

15.4.2 Example: Schlumberger

Depth information is graphically displayed by Schlumberger in the Depth summary listing (Fig. 15.9).

DEPTH SUMMARY LISTING

Date Created: 5-MAY-1999 11:02:25

Depth System Equipment

Depth Measuring Device		Tension Device		Logging Cable	
Type:	IDW-B	Type:	CMTD-B/A	Type:	7-46P
Serial Number:	432	Serial Number:	8732	Serial Number:	8325
Calibration Date:	2-JAN-1999	Calibration Date:	28-DEC-1998	Length:	18750.00 FT
Calibrator Serial Number:	1765	Calibrator Serial Number:	457	Conveyance Method:	Wireline
Calibration Cable Type:	7-46P	Calibration Gain:	1.37	Rig Type:	Land
Wheel Correction 1:	-3	Calibration Offset:	-0.58		
Wheel Correction 2:	-2				

Depth Control Parameters

Log Sequence:	First Log In the Well
Rig Up Length At Surface:	352.00 FT
Rig Up Length At Bottom:	351.00 FT
Rig Up Length Correction:	1.00 FT
Stretch Correction:	7.50 FT
Tool Zero Check At Surface:	-1.50 FT

Depth Control Remarks

1. This is the first log run in this well
2. Speed correction not applied.
3. Up log depth correlated to down log
4.
5.
6.

Figure 15.9

Example of depth summary listing.
By courtesy of SPWLA.

The Schlumberger depth summary listing includes:

- The date and degree of conformance of the wheel master calibration.
- The depth reference used; e.g., kelly-bushing tool zero or tie-in to previous runs.
- The amount of stretch or other environment correction that is applied.
- The closure upon return to surface or depth reference.
- The remarks, especially deviations from standard procedures.

15.4.3 Depth policy documentation and traceability

In addition to the depth information box that needs to belong to any data deliverable, the following information is useful to the oil companies [4]:

- Vendor depth policies need to be public and available to data users.
- Vendors need to perform audits to verify that the depth policy is understood by the employees and fully adhered to.

15.5 RECONCILIATION OF DEPTHS

Several studies indicate that depth can be reasonably managed if vendors provide the relevant information and document their operating procedures [6].

A full-blown case study performed in the North Sea confirms this statement. Fig. 15.10 summarizes the findings. In the left track, differences of 15 m (50 ft) between wireline and LWD depths are observed. Once data is verified and corrected (a stretch correction is also applied to the "driller's depth"), the diverging depths are agreeing within 3 m (10 ft). This amount is still large. Additional work would further refine this agreement.

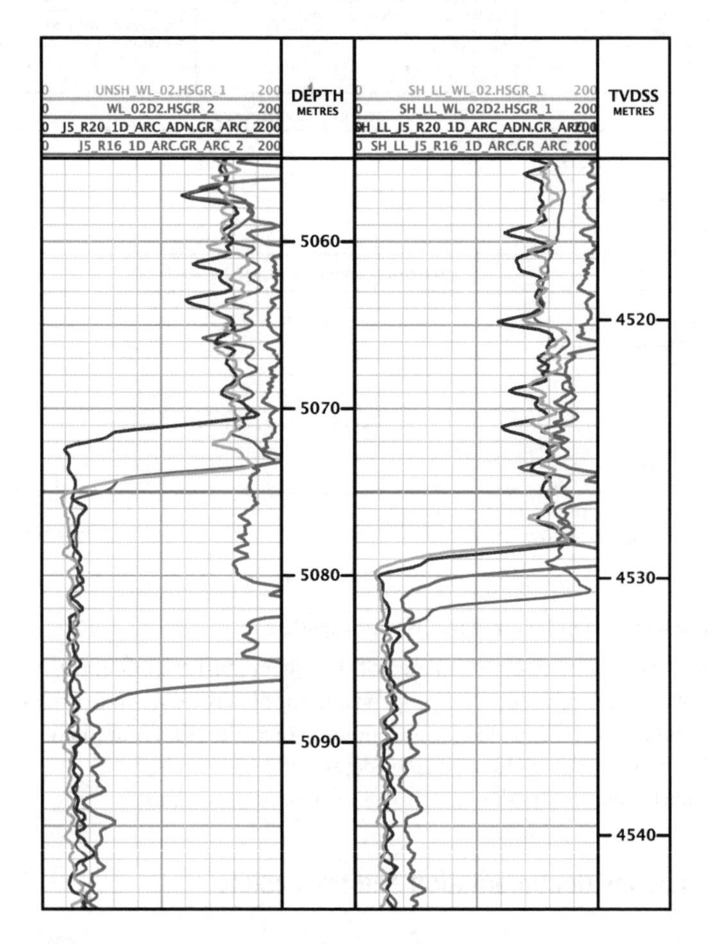

Figure 15.10

Example of depth reconciliation, Kristin field.
By courtesy of SPWLA.

15.6 WIRELINE CREEP

15.6.1 Definition of wireline creep

Absolute depth cannot be known accurately, but it is possible to come back to the same spot in the well through the correlation process. It consists in running two logs over the same interval, recognize some features on these two logs and shifting one log to the other one, the latter being called the reference log.

The correlation of two logs is well understood and is performed with success in most cases, especially when the two logs are run at the same speed. This is the first step taken before taking stationary pressure measurements, collecting sidewall cores and perforating. The following step, stopping the down-hole equipment to complete the operations mentioned before, is not completely controlled [2] and may cause a depth mismatch, causing dry tests, misinterpretation of the samples in thinly laminated formations or off-depth perforations. Taking fluid and core samples or perforating at the wrong depth may result.

Creep is an effect observed in wireline conveyance. It is due to the fact that, when the winch, at surface, is stopped, the down-hole equipment keeps moving before eventually stopping. The distance covered by the down-hole equipment is called creep.

15.6.2 Causes of creep

When the winch is stopped, this event is not instantaneously recognized by the down-hole equipment. The information is propagated to the tool string with a limited velocity. For a 24,000 ft well, it takes about 4 s between the time the winch is stopped and the time the tool string starts slowing down. It takes additional time for the tool to come to a complete stop. This first cause of creep is called cable delay. From the moment the tool starts slowing down, its velocity decays in a quasi-exponential way, with a time constant depending on depth, mud density and mud viscosity.

In less than perfect borehole conditions, rough holes, presence of mud cake, complex trajectory, the tool string is subjected to frequent accelerations and decelerations. This adds an unpredictable component to creep, stick and slip.

15.6.3 Modeling creep

In Fig. 15.11 the amount of creep, that is, the creep distance, is proportional to the area between the curves of winch velocity and tool velocity versus time. Winch and tool depths are assumed to be identical at the start of the plot, and the area between the curves is the integral of the difference in velocity versus time.

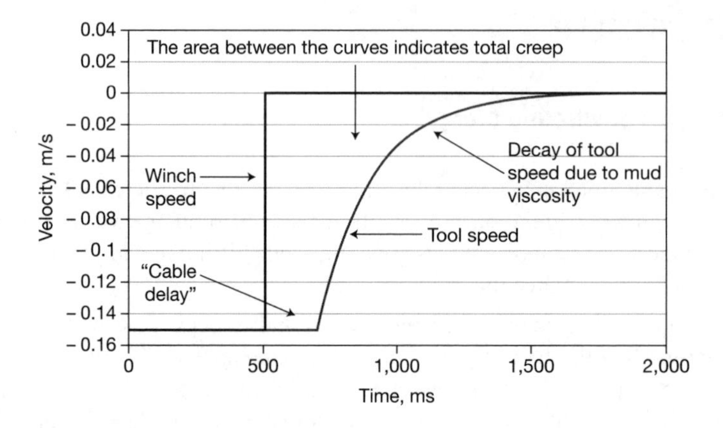

Figure 15.11

Model of creep including cable delay.

The area between the two curves corresponds to total creep. The winch stops at 500 ms.

By courtesy of SPE.

15.6.4 Confirmation of creep

Effects of creep have been confirmed independently by three methods:

- By monitoring the signal detected by the tool after stopping as it approaches a known stimulus (casing collar or radioactive pip tag).
- By monitoring an accelerometer and integrating its measurement to velocity, then to distance.
- By surveying stationary events with subsequent logs. For instance by running a cased-hole imaging log after perforation.

15.6.5 Practical examples

A real example (Fig. 15.12) is extracted from reference [2]. A well documented example considers a creep of 1.7 m (approximately 6 ft).

It is interesting to note that some examples display a negative creep, putting the logging tool string deeper than the depth indicated at surface.

15.6.6 Effects in cased-hole

Reference [7] indicates that wireline perforations are systematically too shallow by a significant amount. Reference [2] gathers data from 52 perforation zones. It yields an average of 0.05 m and a standard deviation of 0.24 m. It is obvious that additional studies and further analysis are needed.

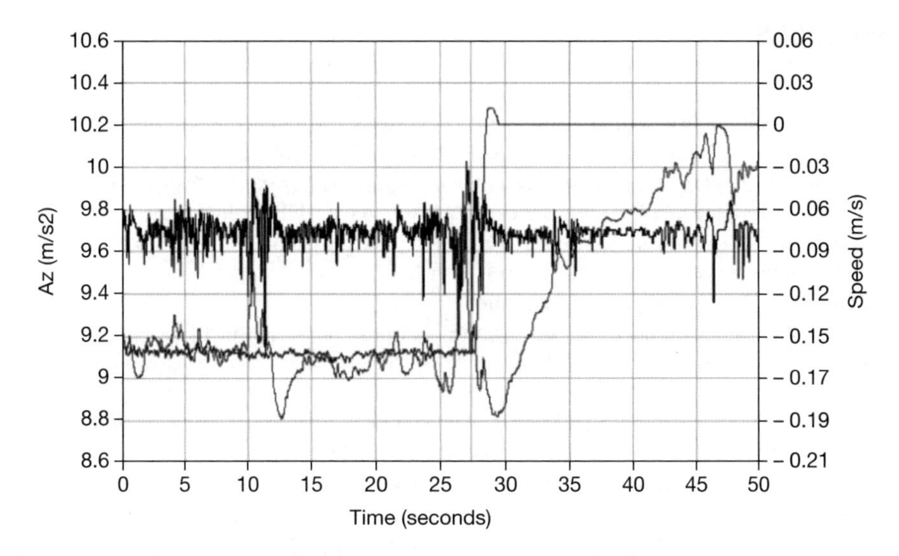

Figure 15.12

Creep monitoring on a real example.
The tool is still moving 20 s after the winch stops.
In this example, the creep is estimated at -1.686 m.
By courtesy of SPE.

15.7 SUMMARY

- Depth is the most important logging measurement.
- There are different depth measurements. No measurement yields true depth.
- Matching depths (i.e., adjusting one depth measurement to another measurement) creates confusion.
- Vendors need to provide all possible information on the way they measure depth.
- Once all auxiliary information is used, it is often possible to reconcile the different measurements.
- When a tool on wireline is stopped from surface, it is likely to creep for a few more seconds, with effects that need to be carefully quantified.

REFERENCES

[1] Desbrandes R., Schenewerk, P., "Stretch depth correction with logging cables," CWLS, 1991.

[2] Fitzgerald, P., "Wireline creep," paper SPE 118027, Abu Dhabi, 2008.

[3] Kirkman, P. S., "Depth measurements with wireline and MWD logs," SPWLA Norwegian chapter, 1989, reprinted in SPE reprint series 40 "Measurement while Drilling," 1995, pp. 27-33.

[4] Loermans, T., Kimminau, S., Bolt, H., "On the quest for depth," SPWLA 1999, 40th SPWLA annual logging symposium, Oslo, Norway, 1999.

[5] McGinness, T. E., personal communication.

[6] Pedersen, B. K., Constable, M. V., "Operational procedures and methodology for improving LWD and wireline depth control, Kristin field Norwegian Sea," *Petrophysics*, Volume 48, N°2, 2007.

[7] Pledger, T., *et al.*, "Are wells perforated on depth today?" *World Oil*, January, 2008.

[8] Schlumberger, Abu Dhabi depth conference minutes, 2003.

[9] Smith, D., "Depth Control on Deep Tuscaloosa Wells," MCBU.

[10] Sollie, F., Rodgers, S., "Towards better measurements of logging depth," SPWLA, 1994.

[11] Theys, P., *Log data acquisition and quality control*, Chapter 20 on directional surveys, Editions Technip, 1999.

[12] Williamson, H. S., Wilson, H., "Directional drilling and earth curvature," paper SPE 56013.

[13] Williamson, "Accuracy Prediction for Directional Measurement While Drilling," paper SPE 67616.

[14] www.ngs.noaa.gov/PUBS LIB/gislis96.html

16

Hidden Treasures

Current logging deliverables are often incomplete. The data vendors often argue that oil companies do not know what to do with what is available. They also comment: "If the oil companies want information, we do have it; and we do not have any problem to deliver it."

It is therefore recommended that oil companies ask for anything that is available.

The discovery of information gems is a matter of luck. In an audit, the expert would find a nice and original feature. He would then think: "Why is this feature not delivered all the time with every log?"

The sections in this chapter gather examples found at random. They do not represent an exhaustive list as it is likely that additional excellent quality features are available. Each section is independent one from another. There is no logical link between them.

16.1 CHART BOOKS

Chart books contain a wealth of information [2]. They are provided by logging companies at a simple request. The chart books are not directly used to perform environmental corrections as software is used to complete this dull and boring task. They are more useful in the planning phase of a logging job or as a training tool to get a data user familiar with potentially uncertain data. They may also contain useful information on the operational procedures and on strategies to minimize errors.

16.1.1 Old charts

It is obvious that environmental effects have also some impact on data acquired years or decades ago. These effects need to be corrected for. Where can the relevant charts be found? The Denver Well Logging Society has done an exhaustive work of collecting old charts from 1947 to 1999. Three of the four compact disks of the package are shown in Fig. 16.1.

Figure 16.1

Old chart books created from 1947 to 1999 have been digitally archived.

Old Schlumberger, Halliburton and Western Atlas charts are found on compact disks 4, 5 and 7. Names of extinct companies such as Gearhart, Welex and PGAC can also be recognized.

By courtesy of SPWLA.

16.2 JOB PLANNERS

Logging companies have planning software that is mostly used internally. For instance, a standoff planner for Schlumberger induction tools is available (Fig. 16.2). It enables the optimal selection of standoff size in consideration of a number of variables including hole size and mud type. Generally, logging companies offer little resistance to clients asking for their use. The software may be operated by the logging company representative and not by the oil company person.

Of great interest is the induction tool planner (Fig. 16.3) which informs the user if the data to be acquired has good chances to be subject to low or high uncertainties. Expected inputs are formation resistivity, obtained from a neighbor well, mud resistivity, standoff size and borehole size. From this information, a point is plotted on a two-dimensional diagram.

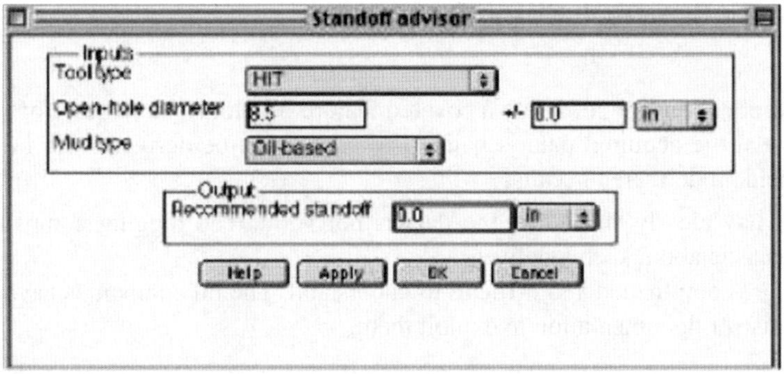

Figure 16.2

Standoff planner.
By courtesy of Schlumberger.

Depending on its location, it is possible to predict if the acquisition job will be successful or not. Note that, in the area entitled "possible large errors on all logs," data is likely to be unusable for quantitative interpretation.

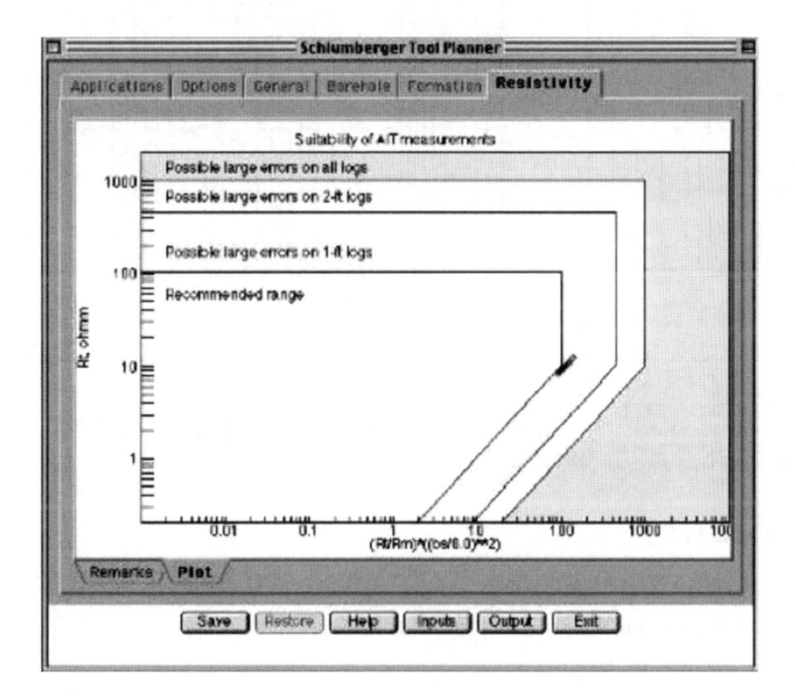

Figure 16.3

Induction log planner.
By courtesy of Schlumberger.

16.3 QUALITY CONTROL CURVES

Quality control curves are definitely a coveted feature of a log. The purpose of these curves is to indicate if the acquired data is valid. The curves can be derived from raw data. The curves are seldom delivered because:

- They may clearly show that the data is not valid. The logging companies are not enthusiastic about such a delivery.
- They are complicated and difficult to understand. The oil companies have seldom the expertise or documentation to exploit them.

The first quality control curve, the density correction, also called $\Delta\rho$, has gained popularity and is used to evaluate mud cake, barite and rugosity effects. As seen in Chapter 13, it can be used to derive a better estimate of the uncertainty on the density itself, ρ_b.

Recently, quality control curves have been displayed on logs as flags, warning the user of any potential trouble. This is the case for the array induction tool, as shown in Fig. 16.4. Fig. 16.5 is a field example underlining the challenging conditions where the tool is run. Fig. 16.6 is an example of quality curves for the tri-axial induction tool.

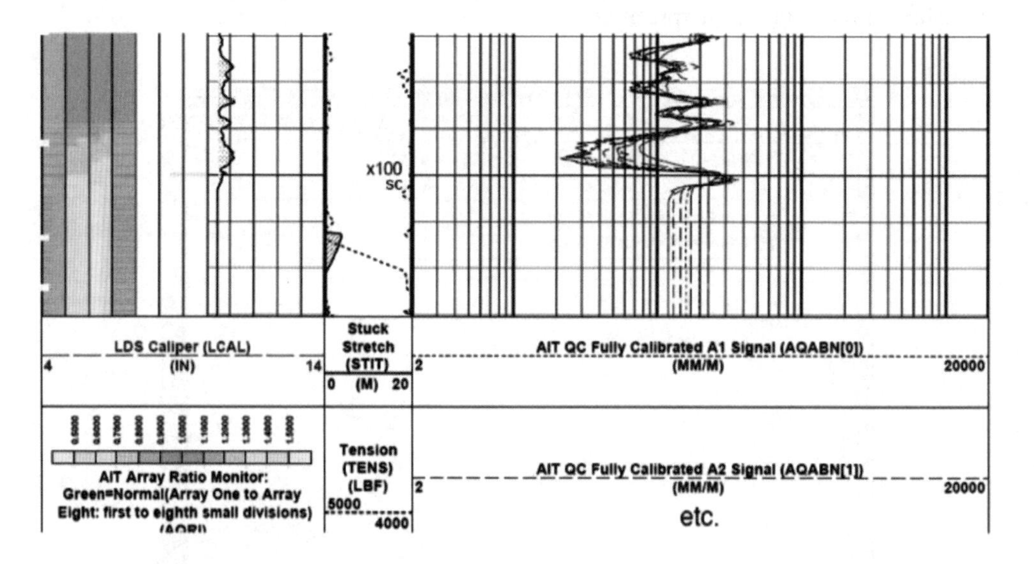

Figure 16.4

Example of QC curves for an induction tool: QC flags are shown on the left side of Track 1. The lighter grey would show as yellow on a color print. Yellow is a warning that not everything is under control.

By courtesy of Schlumberger.

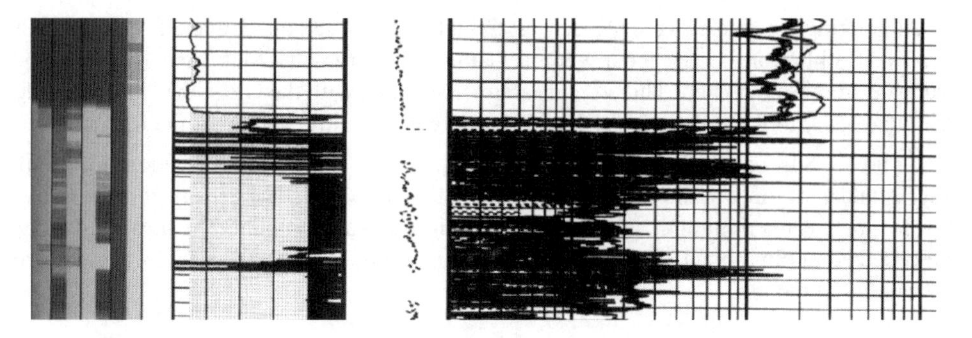

Figure 16.5

This real log is badly affected by poor borehole conditions. The QC curves make the problem quite obvious through grey flags and noisy raw curves.

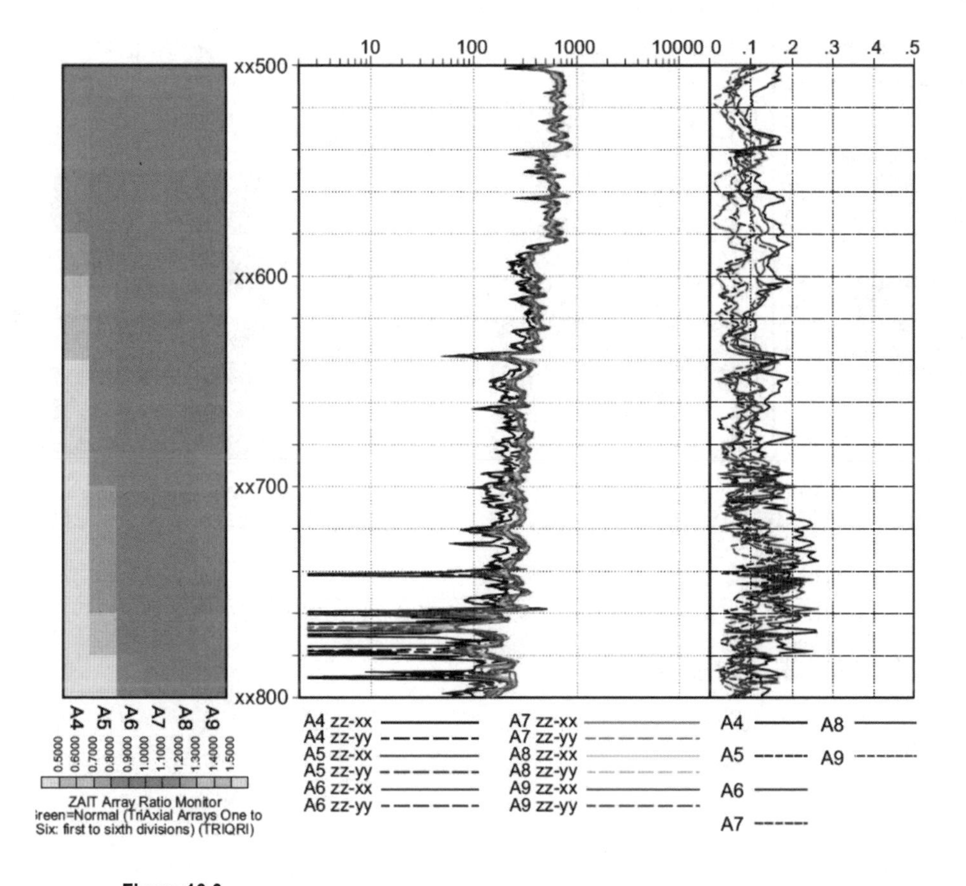

Figure 16.6

Tri-axial induction QC curves.

By courtesy of SPWLA.

All logging companies have QC curves, though they are seldom delivered. Reference [3] displays Halliburton curves for a Natural Gamma Spectrometry tool. Figs 16.7 and 16.8 originate from Baker-Inteq. The QC curves are associated with the ORD, the logging-while-drilling density tool. Explanations of the curves are given at the bottom of the figure.

Figure 16.8 is very instructive. For most density tools, the density correction $\Delta\rho$, should stay around zero most of the time. It is not so for the ORD tool. The density correction may navigate between a minimum, a small positive value, and a maximum. The measured value, the expected minimum and the expected maximum are presented in Track 2. They confirm the validity of the log.

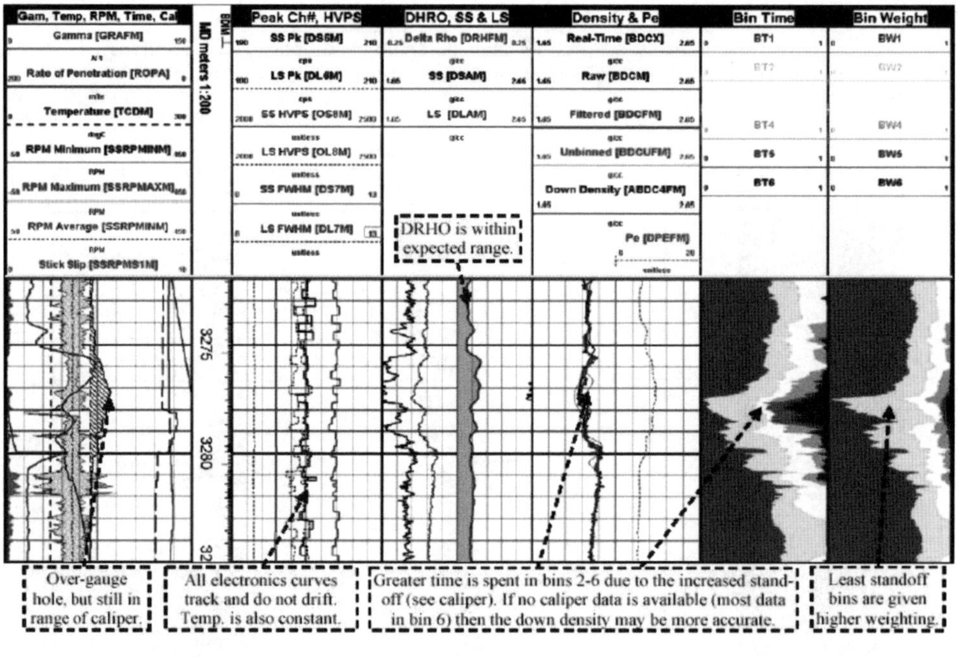

Figure 16.7

ORD QC – Monitoring of the correction curve for the LWD Baker-Inteq density tool.

By courtesy of Baker-Inteq.

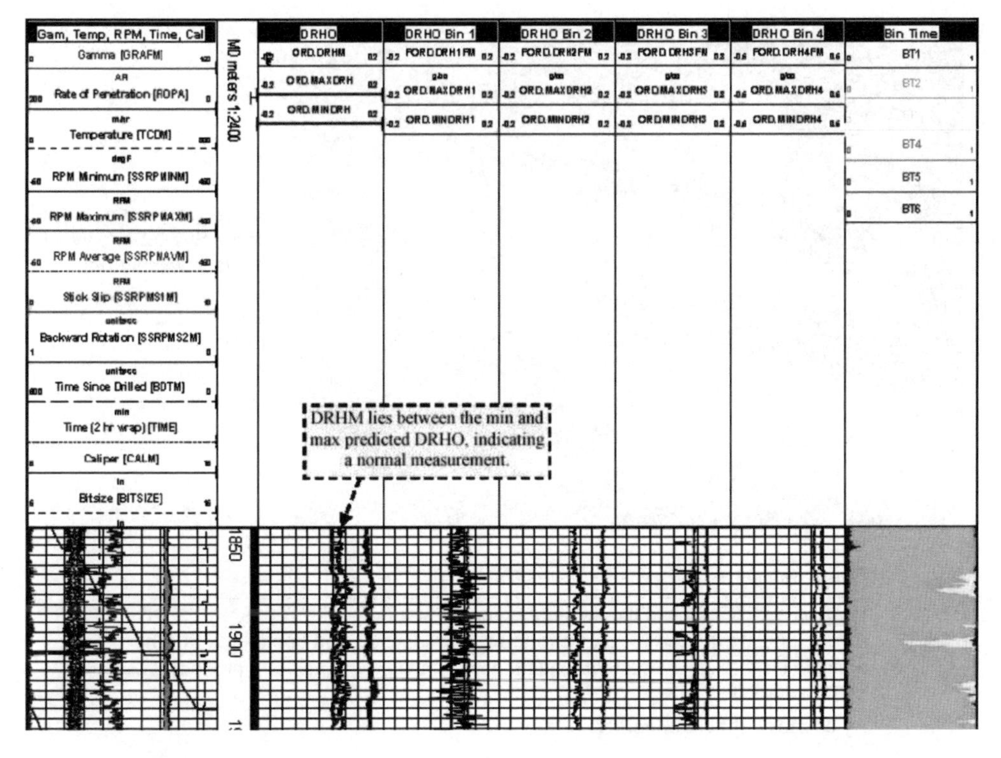

Figure 16.8

Example of predictive drho.
By courtesy of Baker-Inteq.

16.4 DETAILS ON CALIBRATIONS

Calibration control displays have evolved into long series of flags. As indicated in Chapter 8, “OK” flags are not a guarantee that the calibration is correct. The most critical ingredient in a good calibration is a good set of calibrating equipment and rigorous adherence to logging company calibration procedures.

16.4.1 Calibration guides

Calibration guides including pictures of correct calibration setups and a short description of the procedure are publicly distributed. Figs 16.9 and 16.10 are extracted from the Schlumberger calibration guide [1].

Figure 16.9

Example extracted from the Schlumberger calibration guide. The figure clearly shows a standard setup for the calibration of a sonic cement bond log tool. The remarks on the left are recommendations on the correct operating procedures. SFT means safety tube.

By courtesy of Schlumberger.

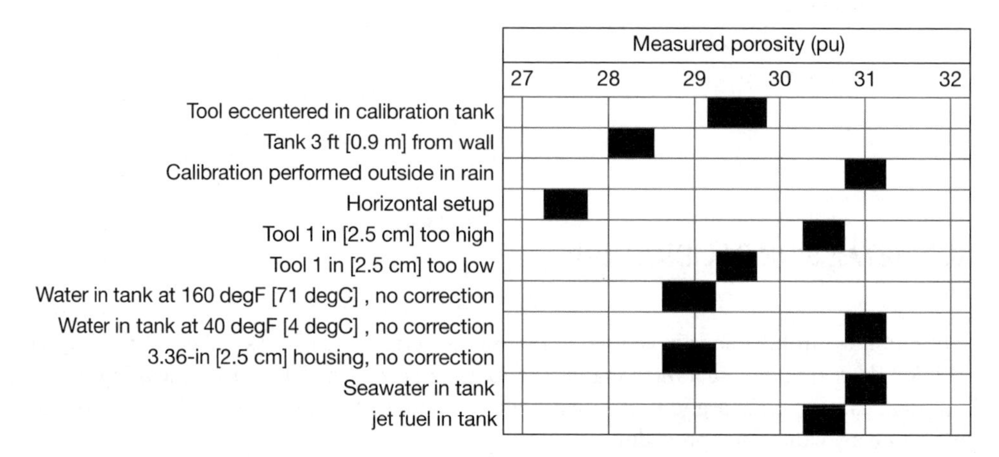

Figure 16.10

Example extracted from the Schlumberger calibration guide: The figure lists the systematic errors introduced by incorrect calibrating procedure. The information can be used to evaluate the uncertainties linked to calibrations. A log with a perfect calibration would read 30 pu.

By courtesy of Schlumberger.

16.4.2 Detailed information on calibration

Logging companies have internal displays that contain much more information than the calibration tail delivered to customers. These displays are often remitted to data users if asked. They are very useful for a detailed uncertainty analysis. The information shown in Fig. 16.11 enables a quantitative analysis of the effect of wear. It includes measurements of the tool in controlled blocks before and after the job.

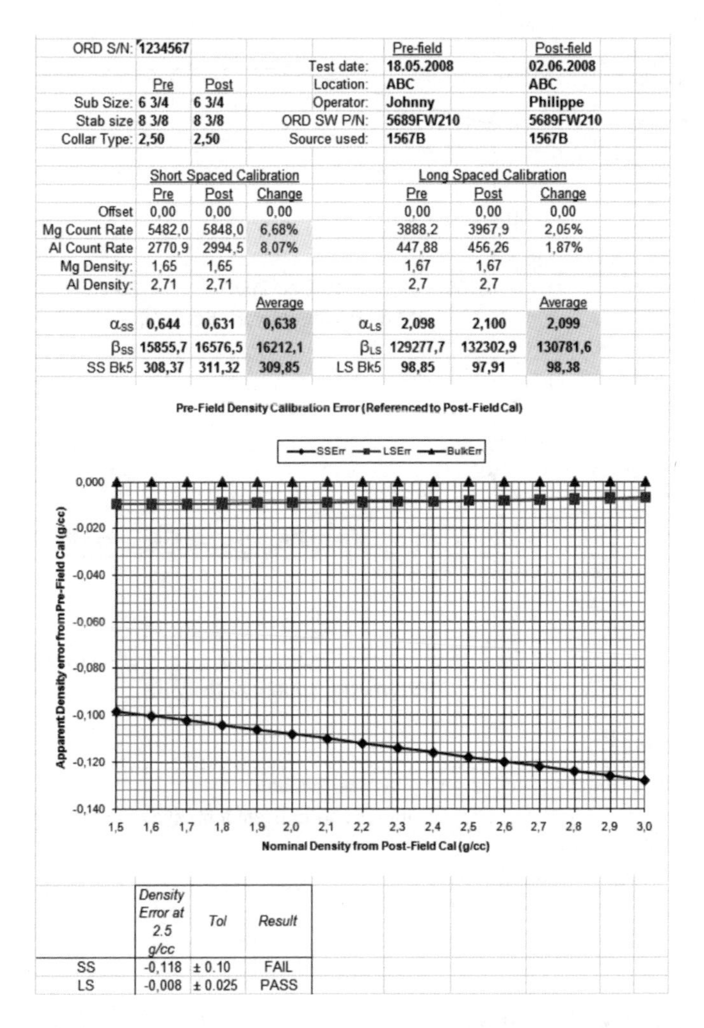

ORD S/N:	1234567			Pre-field	Post-field
			Test date:	18.05.2008	02.06.2008
	Pre	Post	Location:	ABC	ABC
Sub Size:	6 3/4	6 3/4	Operator:	Johnny	Philippe
Stab size	8 3/8	8 3/8	ORD SW P/N:	5689FW210	5689FW210
Collar Type:	2,50	2,50	Source used:	1567B	1567B

	Short Spaced Calibration				Long Spaced Calibration		
	Pre	Post	Change		Pre	Post	Change
Offset	0,00	0,00	0,00		0,00	0,00	0,00
Mg Count Rate	5482,0	5848,0	6,68%		3888,2	3967,9	2,05%
Al Count Rate	2770,9	2994,5	8,07%		447,88	456,26	1,87%
Mg Density:	1,65	1,65			1,67	1,67	
Al Density:	2,71	2,71			2,7	2,7	
			Average				Average
α_{SS}	0,644	0,631	0,638	α_{LS}	2,098	2,100	2,099
β_{SS}	15855,7	16576,5	16212,1	β_{LS}	129277,7	132302,9	130781,6
SS Bk5	308,37	311,32	309,85	LS Bk5	98,85	97,91	98,38

	Density Error at 2.5 g/cc	*Tol*	*Result*
SS	-0,118	± 0.10	FAIL
LS	-0,008	± 0.025	PASS

Figure 16.11

Details on a Baker-Inteq LWD density calibration.

The table compares pre- and post-calibrations. The percentage of shift between the two is indicative of wear.

By courtesy of Baker-Inteq.

16.4.3 Successive calibrations

As explained in reference [3], comparing the calibration coefficients from successive calibrations allows the recognition of anomalous drifts caused by a hardware failure. Fig. 16.12 applies to a density calibration.

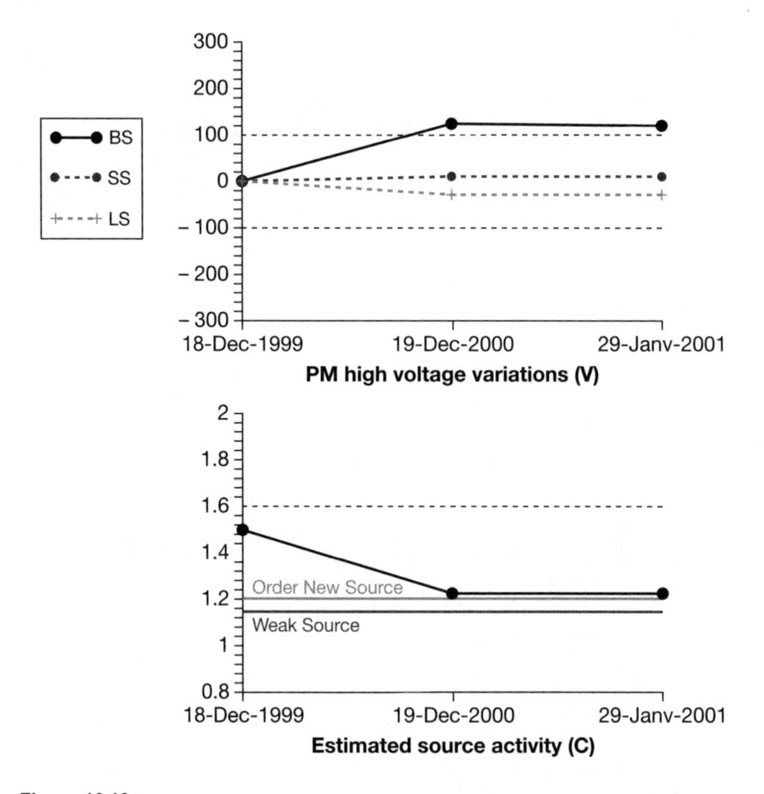

Figure 16.12

Display of the photomultiplier voltage versus time.

An abrupt change would be suspicious and indicate a breakage in the photomultiplier.

By courtesy of Schlumberger.

16.5 INFORMATION ON WEAR

All logging tools do wear when they are run in the hole. A logging interval of 3,600 ft is considered. If it logged with a wireline tool at a speed of 1,800 ft/h, two hours are needed. The tool may also be subjected to substantial wear while running in and running out.

In logging while drilling, the logging time is controlled by the rate of penetration. At 60 ft/h, the survey takes 60 h, 30 times more than wireline. Fig. 16.13 shows $\Delta\rho$ increasing with depth. It is indicative of the continuous wear cumulating as the drill bit cuts the formation.

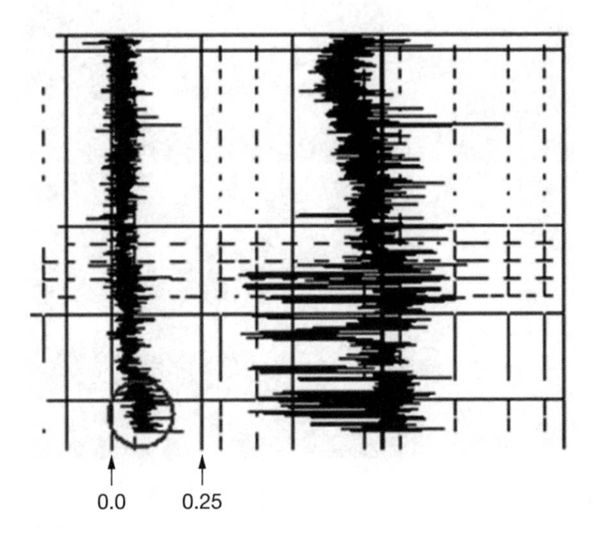

Figure 16.13

Wear observed on the density correction curve.
The depth scale is quite compressed.
The curve on the left is a density correction curve with units in g/cm^3.
The correction starts close to 0.0 g/cm^3 at shallow depths, then creeps slowly as depth increases.

Wear may be detected independently of the direct check of the logs. It can be observed on the logging tool itself. Some logging companies have developed physical hardware features that reveal the presence of wear (Figs 16.14 to 16.16). The information collected from the observation of these indicators is shown in Fig. 16.17.

16.6 SUMMARY

- Logging companies have extensive and excellent documentation whose existence is poorly known by oil companies.
- Logging companies have many quality features developed by the tool designers. These features are not delivered because they are not specifically required by the oil companies.
- Regarding these hidden treasures, data vendors should be more open in their communication with their customers for the benefit of all. Data vendors should also provide the training to the data users so that these features are fully exploited.
- These hidden treasures are useful in the quantification of uncertainties.

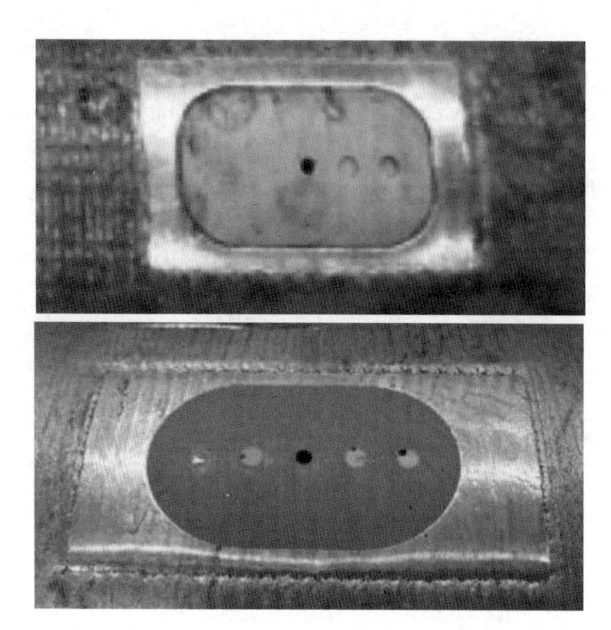

Figure 16.14

Observation of wear on the detectors.

The top picture shows a worn sub. The bottom picture represents a newly refurbished sub.

By courtesy of Baker-Inteq.

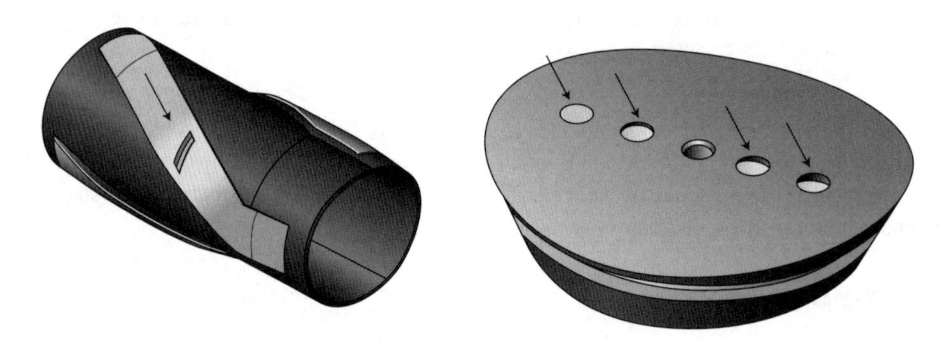

Figure 16.15

Wear indicators on a density LWD ORD tool.

Each hole on the sub has a different depth. Depending on the amount of wear, some holes completely disappear. It is possible to link the thickness of material removed with a shift of porosity.

By courtesy of Baker-Inteq.

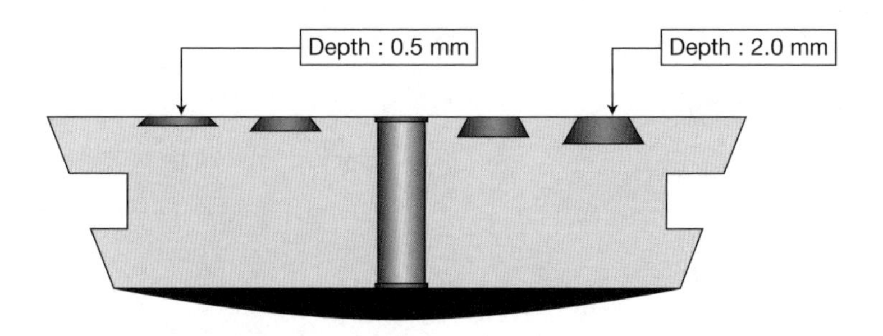

Figure 16.16

Cut view of the wear indicator showing the varying depths of the holes
By courtesy of Baker-Inteq.

Table 16.1 Summary on the findings on the wear indicators
While the pre-calibration and check before the job indicate no wear, some is detected after the job. More than 2.0 mm of sub material has been removed during logging. From the original four holes, none is left after logging. The density has been drastically modified.
By courtesy of Baker-Inteq.

16106B	Pre cal	Stabilizer Wear Indicator:	Not Installed		Levels
		Source Window Wear Indicator:	less than 0.5 mm	4	Holes
		LS Detector Window Wear Indicator:	less than 0.5 mm	4	Holes
	Post cal	Stabilizer Wear Indicator:	Not Installed		Levels
		Source Window Wear Indicator:	more than 2.0 mm	0	Holes
		LS Detector Window Wear Indicator:	more than 2.0 mm	0	Holes

REFERENCES

[1] Schlumberger marketing services, *Schlumberger logging calibration guide*, 08-FE-014, 2008.
[2] Schlumberger, *Chartbook*, 2009.
[3] Theys, P., *Log data acquisition and quality control*, Éditions Technip, 1999.

17

Contribution of the field engineer to the quality of data

Why do you need intelligent persons like us to do that job?
A student from one of the best universities in Europe to a logging company recruiter.

17.1 HISTORICAL CONTEXT

17.1.1 Early field engineers

The history of logging started with people who spent considerable time in the field. The most popular pictures of Conrad and Marcel Schlumberger, and Henri Doll have been taken in field locations (Fig. 17.1). The first employees of the logging companies had assignments alternating between field and research center.

Early recruitment for field assignments targeted non-specialized engineers knowledgeable in all aspects of physics, mechanics, electricity and electronics. They had no difficulty to understand hydrodynamics, mechanics, rheology and three-dimensional electricity.

Considering the intimate interaction of the logging tool, the well and the formation, this was a prerequisite for progress. Field testing started in the Pechelbronn field in France, then was continued in Russia. It is of significance that the spontaneous potential, discovered in 1928 was still being under intense study (Fig. 17.2), essentially in the field, in 1938 [1].

Figure 17.1
Henri Doll calibrating a logging tool in Baku.
By courtesy of Schlumberger.

In these early times, the field engineers did not hesitate to convince the oil company to try many different configurations

ENQUÊTE SUR LA P.S. AUX ETATS-UNIS

(1938)

La SWSC a procédé, parmi ses équipes, à une enquête ayant pour but de recueillir un ensemble d'informations sur le phénomène de la P.S. dans les sondages; informations qui devaient permettre une meilleure compréhension de ce phénomène et, partant, une interprétation plus sûre des diagrammes.

Les réponses des équipes au questionnaire qui leur a été soumis à ce sujet constituent un matériel abondant et intéressant; elles contiennent toutefois peu de choses nouvelles et, d'autre part, ne permettent guère de tirer des conclusions formelles et de trancher définitivement les questions en suspens. Ce fait devait d'ailleurs résulter nécessairement des conditions d'établissement de ces rapports : ceux-ci ne pouvaient en effet avoir pour base une expérimentation faite spécialement en vue de l'étude des problèmes posés, mais devaient s'appuyer presque exclusivement sur le travail courant, travail dont le caractère nécessairement commercial s'accommode mal avec celui d'une recherche expérimentale.

Dans ce qui suit, nous résumerons les renseignements les plus intéressants que l'on peut tirer de l'ensemble de ces rapports concernant les différentes questions qui ont été posées. Nous ferons d'ailleurs précéder quelques-unes de ces questions d'une courte notice théorique et nous terminerons en essayant d'expliquer s'il y a lieu, les particularités observées.

A - Influence de la Nature des Formations

La règle générale est bien connue : les couches perméables, c'est-à-dire formées de grains assez gros pour permettre la circulation des liquides, donnent naissance à des différences de potentiel spontanées nettement différentes de celles qu'on obtient devant les couches imperméables. Dans une série comprenant les deux sortes de formations, par exemple une suite d'argiles et de sables, la valeur de la P.S. est généralement la même en tous points devant les argiles, et cette valeur peut être prise comme valeur de référence, autrement dit comme zéro : les sables se manifesteront par des anomalies de plus ou moins grandes amplitudes.

Cette règle générale peut être théoriquement nuancée, conformément aux

Figure 17.2

Title of field research in French. It can be translated by "Case studies of spontaneous potential in the United States (1938)."

and mud conditions as is shown in Fig. 17.3. The spontaneous potential is run four times, the first time after a full day without any drilling, the second time after the well was swabbed, the third time while flowing oil in the well and a fourth time after a second swabbing operation.

Thanks to this considerable time and reporting on field testing, the early logging curves could be fully understood.

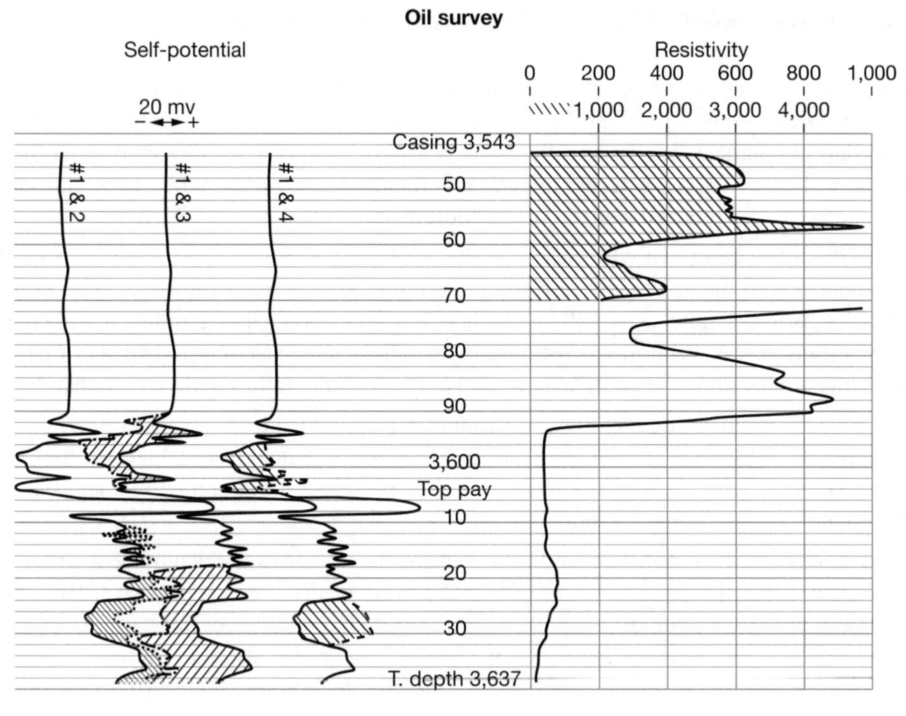

Legend

— #1 Static condition-well idle one day

······· #2 After well was swabbed

—·– #3 While running oil into Hole

· – – #4 After second swab run. Last run

Where curves are not shown they coincide with solid curve #1.

Figure 17.3

Example of field research in 1938.

Several passes are run over the same interval to better understand the spontaneous potential response [1].

17.1.2 The field engineer in the future

What will be the skills required from the field engineer in decades from now? The trend is caricatured by a story modified from reference [2]. *In the future, there will be only one person and one dog in the logging cabin. Computers will run the job. The person will feed the dog and the dog will be trained to bite the person if he has any intention to get close to the computer.*

17.1.3 An intermediate step – Remote support and control

If the role of the field engineer is not yet limited to feed the dog, it is clear that the data vendors plan to assign an onsite engineer with lesser experience. In their vision, the reduced knowledge is compensated by excellent and real-time connectivity to sites located away from the well site with quick access to extensive remote expertise [3] & [5]. These sites may be compared to war rooms distant from the battlefield.

A parallel can be drawn with sport car racing. A wealth of data is transmitted from the car components to the driver, but also to the pit crew. Engine temperatures, engine timing, lambda [1] sensors all indicate overall engine health and can be vital to avoiding a broken car. Individual wheel speed and tire pressure sensors indicate changes in grip. Instead of having a single person handling the data, there are several of them monitoring the task at hand.

In order to optimize logging jobs, war rooms may be added to the expert logger, but are not a replacement thereof. In the sport car, a driver with the qualities of Juan Manuel Fangio [2] or Michael Schumacher [3] is still needed. Racing tracks are designed by men, but oilfields are not.

The development of remote monitoring systems is similar to the deployment of CCTV surveillance. [4] But, the same way criminals make every effort to dodge the surveillance system, reality at the well site or in the wells tends to dodge remote control. Who in the war room can spot excessive stabilizer wear on a LWD density tool just out of a well or the use of LCM [5] that affects the logging curves? Can the remote monitoring confirm that a roughneck is continuously pouring water in the hole at the wellhead while a pressure gauge is run in the hole, or that mud is slowly leaking at surface and thus affects the height of the hydrostatic column in the well. Can anybody in the remote room see that the standoffs on the induction tool are worn-out?

As it takes several days for a team of geoscientists away from the field to perform detective work on limited data sets to understand minute anomalies that represent tens of millions of dollars for the oil company, it is easy to conclude that remote monitoring is not the complete solution. The heterogeneities of nature are such that producing oil and gas is not likely to become a fully robotized process.

17.2 THE HUMAN FACTOR

So it seems that the field engineer is here to stay. Therefore, his behavior needs to be understood.

1. Lambda sensors evaluate the gas/oxygen content in order to optimize the fuel mixture.
2. A sports car driver famous in the 1950s, from Argentina.
3. Another sports car driver, from Austria.
4. CCTV: Close-circuit television.
5. LCM: Lost circulation material.

Unlike the one of machines and computers, the performance of human beings varies much with time. The same person, with the proper training and good general attitude, could be rated 4 one day and 8 another day, on a rating from 0 to 10, 10 being the highest.

This variability is shown on Fig. 17.3 for a runner. The same person can achieve a 52 s-run in his best days, but is limited to 70 s on other days.

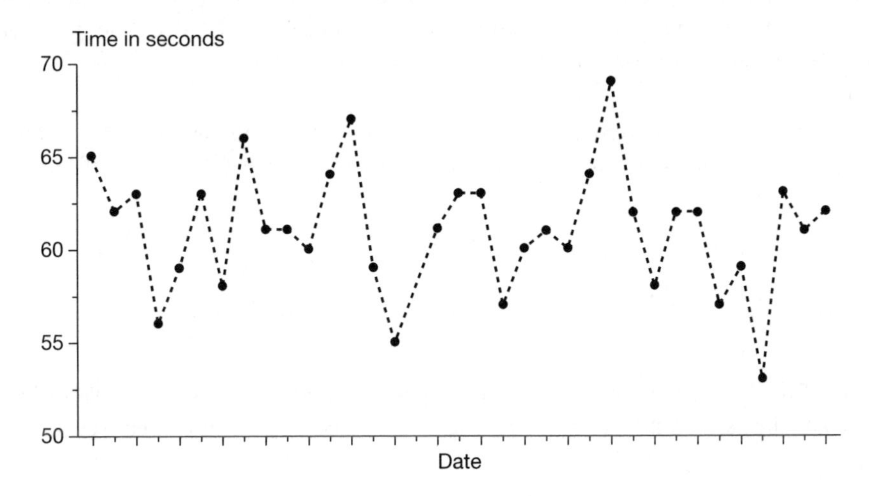

Figure 17.4

Variability of human performance: Example of a runner. The date on the x-axis is in days.

17.2.1 Motivation and performance

A major component of performance is motivation.

Example of bad motivation

An oil company calls a field engineer for a logging job. The engineer has to take a supply boat over rough seas. He is quite sea-sick on the trip. He arrives exhausted on the offshore rig. There is no room for him to rest and the expected timing of the logging job is not clear. He sneaks into the bed of an employee on duty to rest. The communal toilets are clogged and the shower water is either tepid or at freezing temperature. As the engineer finally dives into deep sleep, somebody shakes him to wake him up; the drilling bit will be out of the hole in ten minutes. The disoriented engineer rushes to his unit. The logging program is not clear. Nobody cares about telling the field engineer about the objectives of logging.

It is likely that the data delivery will be a disaster.

Better motivation

The **same** engineer is called later on a second job for **another** oil company. Three days before the job, the exploration manager has called him and has explained the importance of getting excellent data on this specific interval. The engineer is flown on a helicopter. Upon arrival on the rig, he is assigned to an individual room and he is told to rest and get prepared for this critical logging job. He is briefed by the company man, the tool pusher and the mud engineer about what to expect. He is given a clear and realistic estimate of the time when he is going to start logging. Somebody wakes him up gently three hours before the job. The engineer can go to the galley, sip some coffee, then calmly go to his logging unit to prepare the surface equipment. Fresh samples of mud, mud cake and mud filtrate are waiting in the unit with a full report on the mud properties. For a wireline job, a detailed description of the cuttings is also available, so that the engineer can recognize the succession of formations while logging.

This second job has very good chances to be successful even though the same engineer is handling it. When motivated, an engineer tries to use all his capabilities to perform a good job.

17.3 THE MULTIPLE FUNCTIONS OF THE FIELD ENGINEER

The role of the field engineer is to acquire the best possible data considering the drilling of the hole and the use of a given mud. To achieve this objective, he has to wear several hats:

17.3.1 A diplomat

The field engineer may have to convince the rig crew that the tool string configuration, with its standoffs and centralizers is acceptable. Many bad logs have been run because the engineer could not convince the company man to accept these specific pieces of equipment on the tool string.

It may happen that the logging program and procedures have been planned with petrophysicists and geologists, but they are not to the liking of the rig-based personnel. The field engineer must stay firm, but amiable.

Finally, an exercise of tactful negotiation happens when anomalies are observed. They need to be repeated and the local customer representatives should be gently convinced to spend the necessary time to do it.

17.3.2 A safety manager

The field engineer has the critical role to audit his equipment and work environment for any unsafe condition.

17.3.3 An experimentalist, a troubleshooter and an entrepreneur

If the job does not go as expected, the field engineer has to provide solutions. The engineer has to think laterally to solve issues he may have never encountered. Rig personnel pressure needs to be managed. The engineer solves the problem in the most logical way with a calm attitude. Reference [8] relates an incident: "I found myself repairing the truck exhaust pipe with a cup-holder by -45 °C. It was not a piece of cake."

17.3.4 A physicist

The field engineer requires mechanical skills, some minimum knowledge in fluid dynamics, electricity and electronics in addition to an obvious mastering of computer technology.

17.3.5 A reporter

An essential component of the logging job is excellent reporting. Each job is different and these peculiarities need to be described and captured. The style and content of the report are such that, decades later, the data user can fully understand how the job evolved.

17.3.6 The guardian of data

As explained in Chapter 6, there could be a lot of pressure from the logging company management or from the oil company to modify the data. The field engineer has the duty to protect data from being tampered with. He should not be tempted to hide the problems he may have caused, or worse, nudge the data.

17.3.7 A decision maker

The logging engineer is the first person who sees the curves representing the formation. In LWD, he can immediately alert the oil company when a new marker or horizon is traversed by the bit (there is still a delay due to the distance between the bit and the logging sensor). He can alert the oil company if this horizon is reached too early (or too late), enabling a change of plan or trajectory.

With proper connectivity, the information could be sent to a war room, but it is difficult to get the same involvement from a cozy, comfortable room, miles away from the well as from the drill floor shaken but the rotation of the kelly bushing or of the top drive.

17.4 HUMAN ERROR MANAGEMENT

Reference [9] mentions: "When you evaluate the magnitude of the errors that affect logging measurements, you find that the ones induced by the limitations of physics and technology are smaller than the human errors."

Human errors come in two flavors, involuntary, and voluntary, also called blunders. Voluntary human errors are performed to profit the data acquisition company or the data user. They occur in order to hide sloppy procedures, cheat partners, possibly governmental authorities, or benefit individuals. An example is when a person tampers with data to earn a higher bonus. Involuntary human errors, also called blunders, are observed in many fields, much beyond data collection activities, or the oil industry [10].

17.5 BLUNDER MANAGEMENT

As human errors are connected to persons, it is possible to reduce them by shortening the data flow and properly distributing the tasks between the engineer and the surface system.

17.5.1 The field engineer and the computer

The account of the first logging job performed by Henry Doll [6] and by his team can be summarized in a few tasks: *Set the equipment correctly. Adapt to unexpected situations. Use completely transparent algorithms. Verify information. Crosscheck information with cores and cuttings. Monitor depth. Be on the rig floor to see the shape and status of the logging equipment. Report all information and interesting details on the deliverable to client.*

Today, the field engineer acts more as a computer technician than an observer who collects contextual information and delivers it. The field engineer should come back to his roots and be given the means to do so.

This is done partly today with a massive replacement of people by computers. Outside the oil industry, the caricature of this approach is to use barcodes and barcode scanners. They eliminate the need for keyboard entries, which are well-known to be error-prone. It is noticeable that such approach, though most often successful, can sometimes result in generalized errors: when the software is not correct, then all transactions are incorrect. People write software and commit blunders. Entirely bug-free software is unheard of.

17.5.2 Tasks managed by computer technology

Expert information

A lot of the information required on the log originates from experts:

- Location, coordinates, earth magnetic field characteristics, from surveyors,

- Mud information, from mud engineers,
- Cement information, from the cementing company,
- Production data, from production engineers.

This data is complicated and often impossible to understand by the non-expert. Today, most of the information is entered manually by the field engineer. Many mistakes are likely to emerge. This information should be digitally transferred to the logging company. The source of the information should be traced and captured in the log digital record. To facilitate this data collection, digital templates can be defined and standardized by the industry.

Units and definitions

The computer should manage the units (m, ft, psi, kPa, etc.) and do it without any ambiguity. The units and sub-units are to be clearly labeled and shown to avoid confusion.

Examples:

- [NaCl] concentration clearly marked as different from [Cl^-] concentration. The difference could be as much as 65%.
- % weight well differentiated from % volume.
- Gallons [US] are different from Gallons [Imperial].
- ft [US] are different from ft.

Management of names by a database

A way to create disarray is to give little attention to the client, well and field names. These unique data objects should be entered once and never keyed in again.

For instance, a customer order some work on well OCS-G-8932 JC-12, in the East Cameron block 75. Five crews perform logs on the well that has two sidetracks. If the desired attention is not given, this is what could be entered on the log header:

- OCS-G-8932 JC#12 Field: East Cameron block 75
- OCS-G 3289 JC-12 Field: East Cameron block 75
- OCS-G-3289 JC-12 Field: E. Cameron block 75
- OCS-G-3289 JC-12 Field: East Cameron block#75
- OCS G-3289 JC-12 Field: East Cameron block 75.

As a result, five virtual wells are created while reality indicates that there is only one well. A robust database managing the well name alleviates the risk of error. The name of the well is entered once and is only selected by a pull-menu for subsequent usages.

17.5.3 Duties of the engineer

Freed by an improved surface system, the engineer can dedicate his time to:

- Go to the catwalk to check the proper tool configuration (standoff size, use of knuckle joints, etc.).
- Talk to the other contractors to get a feel of the specifics of the well to be surveyed.
- Communicate with the oil company personnel to clarify any possible issue.

17.6 INTEGRITY OR HANDLING VOLUNTARY HUMAN ERRORS

Manipulating data is not new and not reserved to the oilfield. It has been discovered that geniuses have used data nudging to convince other people. Inventors tend to start from a theory (or dogma), then select carefully the data that confirm their theory [7]. One of the fathers of the meter, Pierre Méchain [4] did not report some measurements that were not fitting with the picture he liked.

Therefore, it is essential to preserve all data and make it widely available. The ethical challenge of data suppliers is to spontaneously deliver complicated data that is readable and understandable by the data users.

The more data is delivered, the more likely data quality issues will be found. A manager interviewed by an auditor on the reasons why he was not delivering auxiliary data that was helpful in spotting invalid information, answered: "This auxiliary data may tell the data users that the main data is not correct."

It was first believed that computers would help keep data clean and unbiased in unit equity studies, achieve a degree of impartiality and treat all participants equally. That might have been true with some of the earlier, less sophisticated programs. Today many of the programs used in equity determination are extremely flexible and provide many degrees of freedom. Some companies know how to take advantage of them and are able to optimize the results to their benefit at the expense of the other unit participants.

17.6.1 The field engineer oath

In the same way as future physicians are taking the Hippocratic Oath to practice medicine ethically, the field engineer should be recommended to swear:

- Not to modify data.
- Not to destroy data.
- Not to hide the circumstances of the data acquisition process.

This oath should be taken regardless of the employer and could be managed by professional societies.

17.6.2 Reinforcement in logging companies

Even though the field engineer is put in the right direction by his oath, his attitude regarding ethics needs to be strengthened by the robust stance of the employer. The logging company should protect the engineer from the pressure exerted by internal management and by the oil companies. It should also include sanctions in case the standard is not followed.

Such standards exist [6] but are not given the desired high profile. Logging companies should not hesitate to terminate employees who have a relaxed attitude towards data integrity.

6. Standard 21 at Schlumberger.

17.7 EMPHASIS ON DATA

A final piece of advice closes this chapter. There is definitely some psychological gain to repeatedly emphasize the importance of data.

Some advanced logs (e.g., nuclear magnetic resonance, electromagnetic propagation logs) have been quickly filed without much analysis because they required extra training and expertise. This projects a poor image on the use of data. Also, cement bond logs are often run, delivered and hardly looked at by the customer. The field engineer observing this attitude is not motivated to perform well if he sees that the data acquired with hard work is just not looked at.

A very poor message to pass to a field engineer is: "Anyway, this new data is not that important. This is an old field and we know it well enough." On the contrary, it is better to say: "The data set you acquire is very important and will help us solve all kind of issues."

17.7.1 A log is not only the main pass

Most data users concentrate on the main pass and spend little time on other components (calibration, QC curves, etc.). In order to increase the motivation of the engineer to give attention to auxiliary information, it is necessary to show some interest for other components (Figs 17.5 and 17.6).

Figure 17.5

Oil companies employees are looking at data.

Figure 17.6

Tool designers auditing the deliverables.

17.8 SUMMARY

- The field engineer is an essential contributor to quality data.
- This critical role can receive remote support. An elite group of field engineers is still needed at the well site.
- Blunders happen. Their number can be reduced by a proper distribution of the tasks between the field engineer and the acquisition surface system.
- Voluntary messing around with data also happens. The field engineer oath and the ethical standards published by logging companies should help reduce its occurrence.
- Emphasis on the value of data helps in getting quality data.

REFERENCES

[1] Anonymous, *Evaluation of Spontaneous Potential in the United States*, 1938.

[2] Deffeyes, K. S., *Beyond oil, the view from Hubbert's peak*, p. 26, Hill and Wang, New York, 2005.

[3] Gould, A., "Balancing the interests of consumers and producers," 11th annual Asia oil & gas conference, Kuala Lumpur, 2006.

[4] Guedj, D., *Le Mètre du Monde*, Le Seuil, Paris, 2000.

[5] In-TouchTM, "Satisfied customers rate Schlumberger among the best, Tokyo," ATE World, 2002.

[6] Jost, R., "Le log," *Petrophysics*, July-August, 2006.

[7] Popper, K., *Logik der Forschung*, Springer, 1935. Available in English as *The logic of scientific discovery*, Routledge classics, 1959.

[8] Sébastien, Le blog de Sébastien, http://slb-life.com/?p=648.

[9] Serra, O., *Well Logging and geology,* Editions Technip, 2004.

[10] Theys. P, "Blunder management and traceability, the critical requirements of the longterm value of data," Formation evaluation society of Western Australia, Perth, 2008.

18

Drilling data

Data makes it too complicated so we just do what it takes.

Measuring down-hole shocks is not going to make my drilling tool stronger.

The decisions made from inaccurate measurements are worse than the decisions that would have been made without any data.

Quotes from drilling engineers

18.1 DRILLING DATA, ALSO

Most of the chapters of this book could let the reader think that challenges only relate to petrophysical data. In reality, all kinds of data sometimes fail to be adequate for the users' requirements. This chapter covers some specific aspects of drilling data. There is significant real-time value contained in drilling data for optimizing drilling efficiency, improving operational procedures, avoiding non-productive time (NPT), as well as additional value after drilling for root cause analyses and future well planning.

18.2 HISTORICAL CONTEXT

18.2.1 Surface measurements

Before the advent of down-hole measurements while drilling (MWD), all measurements that helped drilling decisions were collected at surface. They were:

- Depth, block speed and rate of penetration.
- Hook load and weight-on-bit.
- Torque.
- Rotation speed.
- Flow rate.
- Mud density, gas content and composition as measured at the shakers. [1]
- Mud volume.
- Pump pressure.

It is also a standard procedure to have the driller producing a daily report. When well written, these reports contain extremely valuable information. [2]

18.2.2 Enter MWD

MWD adds the knowledge of the down-hole conditions, including:

- Shocks and three-axis vibration.
- Variable rotation speed.
- Axial, torsional and bending forces.
- Annular and bore pressures.
- Flow rate and temperature.

This additional information helps drillers to minimize the risks associated with down-hole vibrations, gas kicks, borehole deterioration, lost circulation, stuck pipe, drill string washouts, and twist-offs. Continuous monitoring gives warnings before the situation takes a catastrophic turn. Some of this information is also quite important to verify the validity of petrophysical data. For instance, excessive vibration and variable sensor rotation speed can explain erroneous formation readings. Variable hookload drill string compression impacts depth accuracy.

The accuracy required from a directional sensor depends upon the well design and the relative location of potentially multiple targets. Positional uncertainty becomes critical in proximity to other pre-existing or planned wellbores, but could be less significant when drilling a short lateral section than an extended horizontal wellbore. Directional data, enabling geo-steering, anti-collision and well placement is already reviewed in [5] and there is no need to elaborate further on the essential role of this type of data and on the need to manage the related uncertainty [7].

1. It should be noted that mud data often seems NOT to vary day after day when drilling. LWD engineers may report constant values of mud density and viscosity when, in fact, the mud engineer adds chemicals periodically and only reports the mud properties when they are back at the specified values. Mud properties constantly change as the added chemicals interact with the formation, and the variation in mud properties should be reported.
2. There is an industry standard for the content of such a report, but most companies adopt their own standard.

18.2.3 Drilling isn't getting easier

Well complexity has been covered in Chapter 7. Extending the length of horizontal wellbores, higher temperatures and pressures, drilling in deeper water, and penetrating past depleted reservoirs are just a few of the increasing challenges faced when drilling. As drilling rigs, tubular equipment and hydraulic systems are pushed to their limits, it becomes ever more important to have good quality drilling data, together with appropriate data flows and associated data processing, in order to enable correct time-critical decision-making.

Even for simpler operations, more than a 100% improvement in some drilling metrics has been achieved simply by monitoring and responding more effectively to changes in the observed drilling parameters.

18.3 THE SPECIFIC NATURE OF DRILLING DATA

18.3.1 What is measured and observed

While petrophysical information relates to rock properties that are reasonably stable, a driller can change the surface drilling control parameters in an instant and drilling information looks into events that vary considerably with time. Down-hole drill string forces and motions are not always evident from the surface measurements and can also change very quickly. Dynamic processes are under scrutiny. The time frame of drill string shocks, vibrations and whirling motions are a fraction of a s,[3] whereas the rotational speed of a bottom hole assembly may cycle over a number of s. Such dynamic events cannot be adequately described from measurements that are averaged over a few s and, similar to the effect of invasion on petrophysical data where measurements of different geometric volumes is appropriate, so the comparison of forces and motions over different time intervals is appropriate for drilling phenomena.

18.3.2 The time domain

Traditionally, averaged or instantaneous values of drilling parameters are recorded for each depth interval that is penetrated while the drill bit is on bottom. Similar to geological mud logs and other petrophysical data, these drilling measurements are displayed with reference to a monotonically increasing hole depth. As discussed above, drilling is a time-related phenomenon – as recorded by a rig's geolograph – and a drill bit passes by the same depth interval multiple times. It is more common, for instance, for a drill string to become stuck when tripping out with the bit off bottom and the critical information pertinent to such a stuck pipe event will not be present in the traditional on-bottom depth-referenced data.

3. It is reminded that the metric abbreviations, s and h, are used for second and hour.

Fortunately, with the advent of computers that are able to handle two orders of magnitude larger volumes of data, drilling parameters are now frequently more appropriately recorded with a time-reference. The attributes of time-indexed drilling data, however, are different to those of traditional depth-indexed data.

It should be noted that even though they may be nominally sampled at regular intervals, time-indexed data frequently contain occasional irregularities. Another issue arising with time-indexed data is that when one data file is sampled every 5 s it might have values at 5, 10, 15 and 20, etc. s past the minute, whereas another data file that is also sampled every 5 s might be sampled at 2, 7, 12 and 17, etc. s past the minute. Data sampling misalignments must be handled appropriately by time-domain algorithms.

Sometimes drilling parameters are recorded at one-minute intervals. Drilling operations, however, can change multiple times in one minute and important drilling events (such as the maximum pull applied to a stuck drill string) may last for only a short instant. A minimum of frequency of 5-s averages of all the standard surface parameters (bit depth, hole depth, block height, block velocity, penetration rate, hookload, weight-on-bit, torque, rotation speed, flow rate, drilling fluid pit volumes, standpipe pressure and drilling fluid gas content) are required for a basic understanding of the drilling process.

Drilling crews are accustomed to coordinating the various depth sensors on a rig so that the driller, mud logger and MWD unit align their depth measurements. The time-indexed data recorded by various contractors, however, is frequently recorded with times offset from one another. Drilling data providers should synchronize their clocks (for both surface and down-hole data) as well as their depth measurements.

18.3.3 Lack of standards

Unlike petrophysical parameters that are recorded with common engineering units, at appropriate sampling rates and with adequate resolution, drilling parameters are recorded in a variety of engineering units, and there are no industry data standards. [4]

An absence of drilling data standards and appropriate data analysis systems are fundamental impediments to drilling efficiency. If the required parameters are not explicitly described, there is practically no chance that they will be delivered. Without an explicit requirement to record data digitally, some drilling contractors assert that they meet their contractual digital data obligations when they simply present drilling data on a digital display.

Oil companies seldom know all the parameters that are available from a particular data vendor who in turn may not be aware of the user's needs. Both parties often lack an awareness of the full value of information contained in the drilling data and, in some cases, neither party is able to appropriately handle or analyze the time-indexed data.

The mud loggers are often responsible for collecting the drilling data which they then use to make calculations of flow rate and mechanical specific energy. The oil company geologi-

4. For instance, fluid volumes are often measured in barrels, cubic feet, cubic meters, liters or gallons (US or UK), and units such as kilodecaN (10^4 N) are sometimes used for easier comparison to imperial units.

cal department, therefore, often specifies the drilling data requirements. This can lead to conflicts related to the limited real-time data bandwidth available from down-hole MWD tools when timely updates of down-hole pressure and vibration data are required in addition to the various petrophysical parameters. Petrophysical frequency requirements change with relative dip. When a drill string becomes stuck or an MWD tool fails, however, there are no more real-time data and drilling operations incur significant additional costs.

All the required drilling parameters are usually available and often recorded at the time of drilling but, due to the large volume of time-indexed data, unless they are specifically requested, the time-indexed data are often discarded soon after the drilling concludes.

A variety of different conventions are used for recording time. Sometimes time is recorded in a 12-h format. Seconds from a time reference is another common time format, but often the time reference and time zone are not included with the digital data. The date format similarly varies (e.g. month-day-year or day-month-year). The required time and date formats should be specified for time-indexed data. [5]

The maximum and minimum hookload, torque and standpipe statistics also can be most informative. For a more complete understanding of the drilling process, a detailed description of the drilling assembly, down-hole pressures, vibration variation in down-hole rotation speed and angular acceleration (stick-slip) measurements, fluid properties (both after **and before** re-conditioning), borehole trajectory and stratigraphic information are required. Other data such as drilling assembly make-up torques and circulating system valve settings should also be recorded.

There are different technical requirements from drilling data for each stage of a drilling operation as various geological strata are penetrated with their associated drilling characteristics.

Hole cleaning is always a necessity, but a vertical hole section is generally less difficult to clean than either a 45° tangent section or a horizontal extended reach wellbore where more accurate down-hole pressure and/or surface torque and drag measurements may be critical for preventing the loss of a bottom hole assembly.

The pressure kick tolerance when drilling out of casing is generally greater than at the end of a hole section where more accurate down-hole pressure measurements may be required to prevent fluid influxes, kicks and lost circulation. Down-hole pressure data can also be most useful for ensuring that the borehole is being opened to the required diameter during under-reaming operations.

Not only is it important to ensure that the sensors used are able to deliver the required data accuracy, but the quality of the actual measurements must also be assured.

18.3.4 Quality control and quality assurance

Automated digital data quality assurance is available. Such systems, however, are generally better suited for meta-data [6] completeness and consistency rather than the accuracy of engi-

5. The time format chaos is managed by an ISO standard: ISO 8601. Time is given as YYYY-MM-DD-hh:mm:ss.

6. Meta-data: Data about data.

neering data for which the responsibility of quality assurance traditionally falls onto the mud logger or MWD engineer. Mud loggers and MWD engineers are not usually responsible for the more complex torque, drag and hydraulics analyses that require accurate measurements and so generally are not necessarily best qualified to assess the data quality. It is generally preferable if persons other than those responsible for acquiring the data assure their quality in order to avoid any possible conflict of interest.

With regard to the meta-data, from one bit run to another, the number of down-hole drilling parameters available may change but the availability of surface parameters should remain the same. Even though they were available, key drilling parameters may be missing from the digital data files, engineering units are sometimes unspecified or incorrect, sampling rates may be inadequate, data resolution poor, null values truncated, time-index corrupted, and/or the formatting inconsistent or incorrect. It is not uncommon when requesting time-indexed data to receive, instead, a depth-indexed data file with time-stamps of the time at which each depth was penetrated which is inadequate for many purposes. Sometimes the time reference itself may be lacking and thus render a data file useless.

In comparison, the computerized analysis of petrophysical data is a more mature art and the input data requirements are fairly well established. Digital petrophysical data files are generally formatted with the required parameters. Sensors can still malfunction and there remains a need to verify the data accuracy, and for petrophysical data there are procedures for assuring their quality that include primary calibrations, monthly re-calibrations, pre-job and post-job operational checks, and repeatability verifications as seen in other chapters. These fundamental quality assurance procedures exist for directional sensors but are currently lacking for most other drilling characterization sensors. There is considerable room for improvement for petrophysical data as explained in other chapters, but, when it comes to drilling data, the foundations for quality assurance are yet to be established.

Difficulty to repeat the measurement

One fundamental difference between petrophysical and drilling data is that a petrophysical sensor may pass across a formation multiple times where it should provide a repeatable measurement. The drilling response (e.g. torque) while drilling through a particular formation with a specific rotation speed and bit force, however, only ever occurs once and so cannot be verified in the same manner. The forces when tripping a bottom hole assembly past a particular depth multiple times is similarly highly variable depending on hole condition, drill string movement (axial direction and rotation speed), and fluid buoyancy effects.

For petrophysical data there is typically only a single data vendor contracted to provide any particular measurement on a particular well. For surface drilling parameters, however, there are often as many as three different vendors (drilling contractor, mud logger and MWD contractor) simultaneously measuring nominally the same parameter. Seldom are their various measurements of the same parameter the same. Part of the reason for this discrepancy is that different data providers use sensors with inherently different accuracies. Even with the same sensor types, however, there can still be some significant variation. The main reason is that drilling contracts seldom include any requirement for drilling data accuracy, so traceable procedures for assuring data quality and verifying sensor accuracy have not been developed.

Data formats

WITSML and LAS data formats are the most popular. WITSML is often used for well metadata, but less frequently for time-indexed drilling measurements. The time-indexed and depth-indexed drilling data are commonly presented in LAS or ASCII formats. Typically LAS files display more consistency while ASCII files are formatted differently by each data contractor as well as sometimes by the various individuals within a particular contractor, even on the same rig.

Data corruption

When some data providers merge together, time-indexed data recorded by different sensors into a single digital file, data may become corrupted.

Fig. 18.1 is an example of a corrupted time-indexed file where the date and time are displayed in the first and second columns respectively. The depth is shown in column five, and the block position in column seven.

Line	DATE	TIME	BIT_DEPTH	HOLE_DEPTH	HOLE_TVD	BLOCK_POS	ROP	WOB	HOOK_LOAD
1	07-19-07	23:55:10	4,625.87	4,625.87	4,624.92	38.76	11.56	40.54	50.36
2	07-19-07	23:57:10	4,626.21	4,626.21	4,625.25	38.41	7.76	38.82	52.08
3	07-19-07	23:58:00	4,626.30	4,626.30	4,625.34	38.32	7.76	36.92	53.98
4	07-19-07	23:53:10	4,626.60	4,626.60	4,625.64	38.02	7.76	39.59	51.31
5	07-19-07	00:00:20	4,627.36	4,627.36	4,626.40	37.26	11.97	43.01	47.89
6	07-19-07	00:06:30	4,627.48	4,627.48	4,626.52	37.15	11.97	34.91	55.99
7	07-19-07	00:07:20	4,627.80	4,627.80	4,626.84	36.82	11.97	34.88	56.02
8	07-19-07	00:02:30	4,627.92	4,627.92	4,626.96	36.71	11.97	33.12	57.78
9	07-19-07	00:05:50	4,628.01	4,628.01	4,627.05	36.62	12.46	31.96	58.94
10	07-20-07	00:05:10	4,628.15	4,628.15	4,627.19	36.48	12.46	34.12	56.78
11	07-20-07	00:06:00	4,628.24	4,628.24	4,627.28	36.39	12.46	32.74	58.16
12	07-20-07	00:09:20	4,628.33	4,628.33	4,627.37	36.29	12.46	32.05	58.85
13	07-20-07	00:04:30	4,628.42	4,628.42	4,627.46	36.2	12.46	32.26	58.64
14	07-20-07	00:07:50	4,628.51	4,628.51	4,627.55	36.11	12.46	34.23	56.67
15	07-20-07	00:03:00	4,628.61	4,628.61	4,627.65	36.02	12.46	32.05	58.85
16	07-20-07	00:08:00	4,628.72	4,628.72	4,627.76	35.9	15.48	33.02	57.87
17	07-20-07	00:10:30	4,628.81	4,628.81	4,627.85	35.81	15.48	33.04	57.86
2	07-19-07	23:57:10	4,626.21	4,626.21	4,625.25	38.41	7.76	38.82	52.08

Figure 18.1

Corrupted time-indexed data.
By courtesy of M. Hutchinson.

Whereas the depth (column five, grey) monotonically increases downwards as drilling progressed and block position decreases, the corrupt time values in column three can be seen frequently to reverse (highlighted in grey). This would imply that some deeper depths were drilled before the shallower depths which is physically impossible.

In the same figure, the date incorrectly changes as much as five minutes and ten seconds past midnight and thus falsely indicates a sudden 24-h reversal in time – followed shortly thereafter by a false 24-h advance.

Calibration and corrections

Surface sensors are calibrated and verified to be within specification during manufacture. Once installed on the rig, periodic checks of the accuracy of some sensors are made but the formal documentation and traceability are not usually available. Operational checks of MWD sensors are also common. These checks are generally just functional or qualitative, however, and not a quantitative verification of accuracy. The related documents are seldom included in the deliverables.[7]

Surface weight-on-bit sensor designs are often indirect measurements and susceptible to sheave friction and other environmental effects. The surface weight-on-bit measurements are tared in order to compensate for changing drill string length and buoyancy effects, but a simple tared offset does not account for changes in sensor sensitivity or drifts in the accuracy of the measurement over its full range.

Some down-hole weight and torque sensors are both pressure and temperature compensated. Other sensor designs are uncompensated and the resulting inaccuracy renders their down-hole weight and torque data unreliable. Taring can improve the accuracy of down-hole measurements, but taring does not account for the frequent changes in down-hole pressure that occur while drilling. Sometimes even after taring the surface torque appears to be less than the down-hole torque and the surface weight-on-bit less than the down-hole weight-on-bit, which is incorrect.

18.4 DATA PROCESSING

There are various issues specific to the processing of time-indexed drilling data, beyond the acquisition of correct input data. Petrophysical measurements are made from sensors located at different points along a drilling assembly or string of wireline tools. Petrophysical data processing accounts for the relative displacement between sensors and appropriately aligns the measurements onto a common depth-index. This is still subject to the inaccuracies inherent in the depth measurement. Drilling sensors are similarly located at different points along a drilling assembly, but when events happen, the responses occur simultaneously and should not be shifted onto a common depth-reference.

The fundamental calculation of penetration rate is traditionally made from depth-indexed data by dividing a change in depth by the associated change in time. This simple algorithm is

7. Down-hole directional sensors are always calibrated and, when deemed necessary, the surveys are additionally corrected for drill string sag and/or local dynamic changes in the earth's reference magnetic field.

adequate when the bit is continuously on bottom, but if during the time interval the bit is pulled off bottom (e.g. for drill pipe connections), [8] then the traditional calculation is incorrect as the change in depth will actually have been drilled over a shorter time interval with a higher penetration rate. The correct calculation requires a more appropriate time-referenced algorithm.

Even the measurement at surface of the bit depth used to calculate penetration rate is a dynamic phenomenon influenced by changing flow rate, pipe compression (hookload) and rig heave when the drilling rig is a floater [4].

It is also not uncommon for the signal from a digital surface rotation sensor to be sent to an analog data acquisition system which then can introduce 10% errors in the subsequent digital data when compared to the actual rotation speeds measured down-hole by an MWD sensor.

18.4.1 Example of incorrect processing

Figure 18.2 shows the true vertical depth (TVD) in track one, hookload in track two, down-hole annular pressure (PWD) in track four, and the down-hole equivalent mud weight (EMW) in track five which is computed by dividing the annular pressure by vertical depth.

The annular pressure can be seen to change frequently over the entire time interval, but inappropriate depth-indexed processing computes an incorrect constant equivalent mud weight between 10:00 and 10:30.

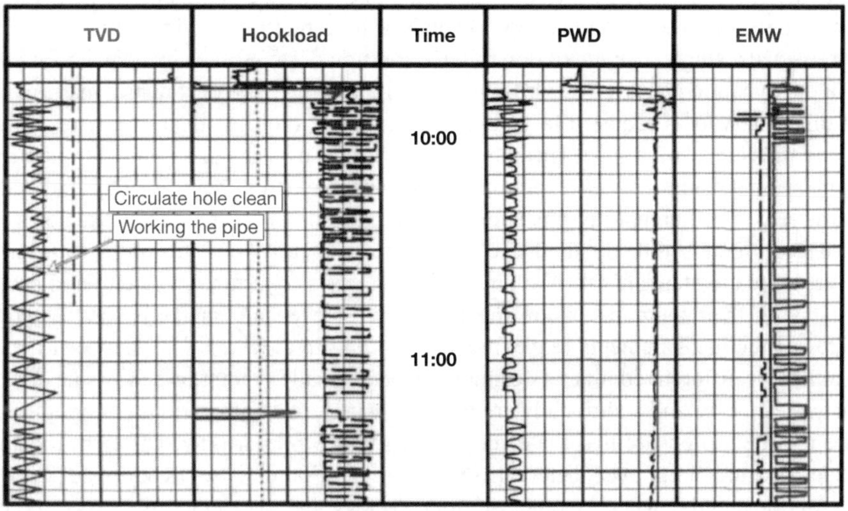

Figure 18.2

Incorrect time-indexed data-processing.
By courtesy of M. Hutchinson.

8. Figure 14.13 is informative on the fraction of time during which the bit is pulled off.

18.4.2 Sampling, processing and statistics

Petrophysical sensors are sensitive to the spatial volume of geological formations and borehole that they pass. Their measurements are influenced by the physical dimensions of the sensor and have spatial attributes of depth of investigation and vertical resolution. For petrophysical data measured while drilling the sampling rate and velocity of the sensor (penetration rate or logging speed) and lateral bending vibration influence the vertical resolution and precision of the data [1].

For drilling parameters it is temporal factors such as the frequency content of the raw signal, the frequency with which data are sampled, the measured statistics (e.g., linear average, root mean square, maximum or minimum), and the time interval over which statistics are acquired that are most significant – especially for dynamic drilling phenomena. There is no point in sampling data every ms if the analog sensor response is passed through a 1-Hz low-pass filter. The time interval over which statistics are measured must also be appropriate for the drilling phenomenon that is being described. For instance, if torsional stick-slip has a period of 7 s, then measuring the difference between maximum and minimum rotation speed over 1/2 s is inappropriate. The sampling period of average forces should also be longer than the dynamic variation, and conversely when measuring the maximum bit force, the sensor frequency response and sampling rate should be faster than the duration of the event.

The invasion of the borehole fluid into a formation requires multiple resistivity measurements, each with a different depth of investigation, to determine the true resistivity of that formation. Drill strings frequently change their vibration state, and cuttings beds routinely build and then collapse. As an analogy to depth of investigation and invasion for resistivity measurements, a single average of down-hole vibration or pressure is inadequate for characterizing the dynamic nature of the drilling process. A comparison of statistics made over different time intervals is required to give an indication of the time-distribution of vibration, force and pressure, and a more meaningful description of a dynamic and complex hydro-mechanical drilling process. For petrophysical measurements there are physical industry standards (e.g. API gamma-ray pits and elemental densities) so that one contractor's measurement corresponds directly to that of another contractor. [9] For drilling measurements, however, there are standards for the physical parameters (e.g. force, velocity and acceleration) but with their temporal nature there are few or even no standards for the temporal attributes of the measurements (e.g., frequency), and direct comparison between measurements made by different contractors cannot always be made. Vibration sensors have various frequency responses and additionally are physically mounted with various damping schemes that are required to ensure that the electronics survive while drilling. Different tools make vibration measurements over different time intervals. Some sensors count shocks (over a variety of thresholds) whereas other sensors average the accelerations. Users should be aware of which sensors types are being used and which statistics being measured.

---

9. In spite of these common standards, it is possible to observe discrepancies between measurements acquired through similar physics and design.

18.4.3 Graphical data presentation

Petrophysical measurements are still generally presented in the same format that data were traditionally recorded using cast-iron optical film recorders within the constraints of their construction, and this makes it easier for drillers, geologists and petrophysicists alike to interpret and use the data. Drilling parameters, however, are currently recorded by the different data collection companies in a seemingly random variety of graphical formats, making data interpretation more difficult.

A standard manner in which drilling data are presented graphically, with similar parameters measured both at the surface and down-hole displayed together (Fig. 18.3, below), is a simple way to improve the ability of the various users of drilling data to recognize drilling events and to anticipate drilling problems before they become more serious [2].

18.5 COST OF INFERIOR QUALITY

Some drilling programs have budgets that anticipate as much as 25% non-productive time (NPT), but even so the actual drilling costs frequently exceed the budget by a greater percentage.

For petrophysical data, an inaccuracy of a few percent can be significant and data quality is assured accordingly. When drilling with a down-hole motor, pressure anomalies may be attributed erroneously to changes in the motor torque when drilling through different lithologies. Similarly, surface torque aberrations related to drilling dysfunctions that are less than a rig's torque capacity are also frequently simply attributed to geological variation and so go unheeded.

Drilling inefficiencies sometimes result in more than a 50% change in a measurement and in such situations, even with inaccurate data, a driller is able to recognize symptoms and react. Many costly drilling inefficiencies, however, only induce a change of less than 20% in a measured parameter. When such changes are less than the precision of the measurement, they often are not recognized by a driller using the instantaneous surface instrumentation and also go unheeded.

Whenever calibrated MWD drilling sensors are available, it is possible to use these tools to verify rig sensor measurements. For instance, on one drilling operation where multiple drill string connection failures were experienced, an MWD weight and torque sensor was requested to try and better understand the dynamic down-hole drilling forces that might have caused the connection failures. The calibrated MWD torque sensor revealed that the rig torque measurement was seriously out of calibration, and it was the rig that was actually applying excessive torque to the connections at surface during their make-up.

On another occasion the simple failure to display down-hole differential pressure resulted in an unopened under-reamer going undetected. The casing later became stuck in the under-gauge hole section, borehole instability was falsely diagnosed, and needless side-tracking resulted.

18.5.1 Example of an incorrect decision as a result of poor fluid data and inadequate data flow

Figure 18.3, below, shows, on the right, the standpipe pressure (black), the down-hole bore pressure (grey on the right of the black curve) and the down-hole circulating density (grey on the left of the black curve) in track eight. The standpipe pressure started to decrease on August 24th at 14:48, as a fluid with unreported rheological properties was being pumped down-hole. The down-hole pressures, however, did not start to decrease until six minutes later at 14:54 after the changed fluid properties reached the down-hole MWD sensors.

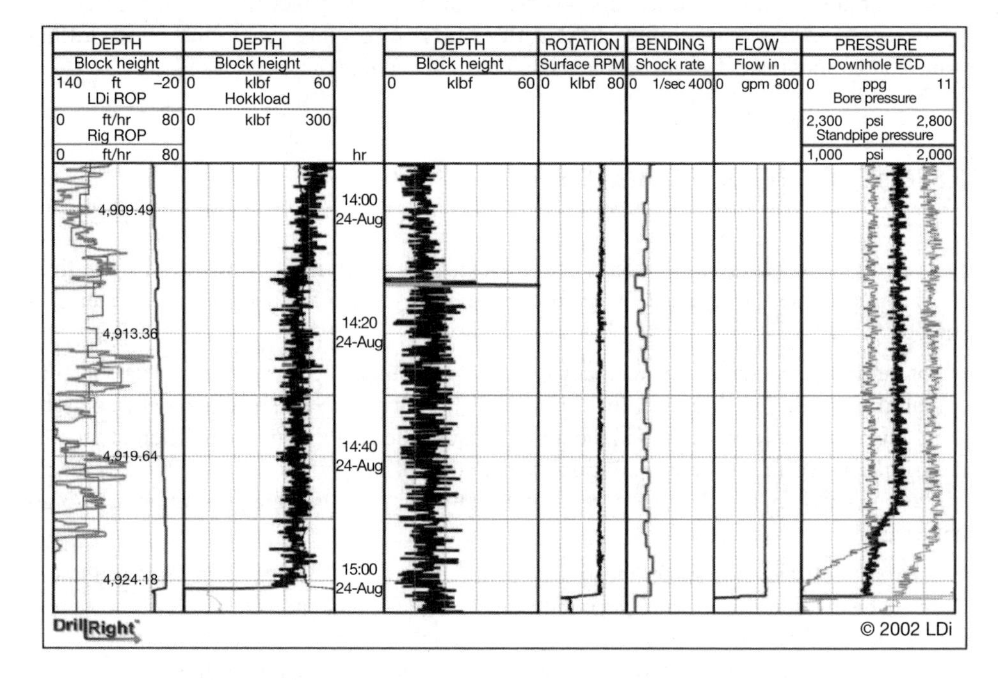

Figure 18.3

Unnecessary trip due to unrecorded change in fluid properties.
By courtesy of M. Hutchinson.

In the absence of accurate drilling fluid properties and the down-hole data not being readily available to the decision-makers, the cause of the standpipe pressure drop was interpreted as a drill string washout and the drill string tripped out of the hole. In fact, there was no washout. A washout would have caused the down-hole and surface pressures to start decreasing simultaneously. During the subsequent tripping operation a gas influx was inadvertently swabbed into the borehole and the stripping procedure left rubber material in the drilling fluid. When testing the well productivity the flow-testing equipment became plugged with the rubber material which ultimately required a new re-entry project – all as a result of poor quality drilling fluids data and down-hole data not being made available to the driller in a timely fashion.

18.5.2 Non-productive time (NPT) and Invisible lost time (ILT)

Drilling improvement initiatives [10] have shown increases in efficiency from 100% to 400% using various traditional drilling performance metrics. If a well can be drilled in half the time and budget [11], this means that the real opportunity is not just a few percentage points of NPT reduction. Oil companies should reference their drilling performance to the variation in historical performance and technical limits (maximum historical performance) as opposed to averages of previous performance [6]. The compression of NPT is important, but so too is the compression of invisible loss time [12]. Better quality data will help accomplish both objectives.

18.6 IMPROVING COMMUNICATION

The oilfield drilling industry has a rich heritage of hydro-mechanical devices stemming back from cable tool rigs with rock chisels and bailers, but tool manufacturers and drilling departments often lack an appropriate data culture. Similar comments to those written as quotes at the beginning of this chapter are still prevalent today.

Contractors acquiring drilling measurements generally have the ability to provide adequate data quality for many purposes when requested, but in the absence of clear requirements and procedures, it should be expected that data quality lacks consistency.

Drilling departments should start the dialogue with their data providers and get intimately involved in the procurement of which drilling data are required for each stage of a well program. Data specifications should include accuracy, resolution, precision and frequency of sampling (especially for real-time MWD data), engineering units, to whom the data need to be presented, in what graphical presentation format, and in what timeframe so that better and timely decision-making is enabled.

Data providers should be required to document the traceability of sensor calibrations and periodically verify their accuracy. Digital data should be archived as soon as it is acquired and the responsibility of assuring adequate data quality should be assigned to persons who are familiar with the use of the data both while drilling as well as for later analysis.

An appropriate methodology for calculating drilling efficiency should be established with a clear definition of which BHA run meta-data statistics are required for assessing drilling performance improvement opportunities. The meta-data requirements should include common definitions for BHA types, steering mechanisms, real-time performance monitoring sensors (e.g. down-hole pressure and vibration), down-hole vibration statistics, vibration isolators, drilling fluid types, performance analysis applications (e.g. torque and drag), drilling failure mechanisms, bit condition, non-productive time, and reasons tripped.

---

10. For instance, FAST drill, by ExxonMobil.
11. Borehole quality, still, should not be put in jeopardy.
12. Invisible lost time (ILT) quantifies the degree of inefficiency that is not apparent using traditional metrics for a well delivery operation. ILT accumulates if rig operations are not performed as efficiently as they could be carried out with currently available technology and best-practice know-how.

18.7 DRILLING DATA IN THE FUTURE

Quality drilling data can be a significant first step towards lowering drilling costs, reducing environmental impact, and in extreme cases saving lives. When combined with appropriate analyses and models, quality data improves our ability to identify the early onset of potential problems while drilling, tripping, running pipe or cementing, and can also help prevent symptoms from developing into more serious events. Examples of the value of quality data are the top-drive control systems that assist with direction control when steering, and that are used to identify and mitigate the destructive rotational stick-slip drilling dysfunction. Such systems have demonstrated significant improvements in penetration rate and bit life that are only possible with accurate data with sufficient resolution that are measured at an adequate frequency.

Significant improvements in drilling performance and rig safety have also been achieved with remote real-time collaboration centers. Nevertheless, due to their remoteness and the inability to make additional contextual rig site observations, they critically depend upon good quality data.

Ultimately, better data quality will enable the automation of more drilling processes with less uncertainty.

Tremendous results have already been achieved by some operators by simply using available data more effectively, and the value of quality drilling data is starting to be appreciated.

An experiment, analogous to the Conoco test wells of the late 80s and early 90s [3] with multiple down-hole drilling sensors run simultaneously in a single bottom hole assembly (and distributed along the drill string) would demonstrate the differences in the quality of measurements provided by different sensors with their respective signal processing schemes. It would also help the industry better understand the influence that doglegs, borehole enlargement, and the location of the sensors in the bottom hole assembly have upon their measurements.

18.8 SUMMARY

- The value extracted from drilling data soared with the advent of down-hole MWD measurements.
- Depth-indexed drilling data have their uses, but even more information and value are contained in the time-indexed drilling data.
- There are no standards for delivering drilling data. This seriously limits its proper use and quality control.
- Inferior drilling data and inadequate analysis can incur exceedingly large costs.
- A large amount of drilling data is acquired, of which only a little is analyzed.

REFERENCES

[1] Hutchinson M.W., *et al.* "Log quality assurance of formation evaluation measurement while drilling data," SPWLA paper HH, Tulsa, 1994.

[2] Hutchinson Mark, *et al.* "Using down hole annular pressure measurements to anticipate drilling problems," SPE paper 49114, New Orleans, 1998.

[3] Hutchinson M.W., "Comparisons of MWD, wireline and core data from a borehole test facility," SPE, paper 22735, Dallas, 1991.

[4] Kirkman, P. S., "Depth measurements with wireline and MWD logs," SPWLA Nowegian chapter, 1989, reprinted in SPE reprint series 40 "Measurement while drilling," 1995, pp. 27-33.

[5] Theys, P., Log *data acquisition and quality control*, chapter 20, Éditions Technip, 1999.

[6] Van Oort, E., Taylor, E., Thonhauser, G., Maida, E., "Real-time rig-activity helps identify and minimize invisible lost time," *World Oil*, 2008.

[7] Williamson, H.S., "Accuracy prediction for directional measurement while drilling," *SPEDE* 221-233, 2000.

19

Coring data

The only sure thing about a core is that it does not belong to the reservoir any longer.
A common ironic quote from petrophysicists.

19.1 MEASUREMENTS ON CORES

With all the complexity and complication found with log data, it could be inferred that cores should be used as a better alternative to logs. In fact, geologists love cores. Cores and core plugs look much more real than curves on paper or stacks of listings. One can touch a core plug, see it and smell it.

But, how is quantitative information derived from core plugs? It comes from measurements performed on core plugs. And measured values are not real values. Any measurement, even on an object as accessible as a core plug, is affected by uncertainties.

19.1.1 Multiple measurements on core plugs

Chuck Neuman [1] decided to perform an interesting experiment. He got core plugs from the same depth and sent them to four laboratories using different methods:

1) Summation of fluids
2) Brine
3) Helium (200-psi overburden)
4) Helium (3,000-psi overburden).

The results are gathered in Table 19.1.

The author computes a standard deviation of 1.02 for the averages (it is computed from the four numbers of the last line). It is also possible to compute the average, the standard deviation and the ratio of standard deviation over average for each individual depth (these numbers have been added to the figures of the original technical paper).

Table 19.1 Core measurements performed with different methods
By courtesy of JPT.

Depth	Porosity (pu)				average	st-dev	(%)
ft	1	2	3	4			
6,023.33	20.20	13.10	15.40	14.60	15.83	3.07	19.4
6,034.42	17.60	18.30	19.00	18.10	18.25	0.58	3.2
6,040.42	11.60	13.80	15.70	14.90	14.00	1.78	12.7
6,061.42	12.00	11.70	14.00	13.60	12.83	1.14	8.9
6,066.33	14.90	11.50	13.70	12.10	13.05	1.54	11.8
6,090.50	11.30	10.90	12.30	11.30	11.45	0.60	5.2
6,109.42	14.90	9.60	10.20	9.30	11.00	2.63	23.9
6,115.42	17.50	15.10	15.50	15.00	15.78	1.17	7.4
6,124.42	8.70	8.10	8.20	7.30	8.08	0.58	7.2
6,148.42	9.30	5.10	1.30	1.20	4.23	3.84	90.9
Average	13.89	11.72	12.53	11.74	12.47	1.02	8.2

The last column displays a fair appraisal of the reproducibility of the core measurements. [1] Large relative uncertainties (up to 90.9%) are observed on the measurements.

19.1.2 Additional issues on core plugs

It is noticeable that the measurement processes in the reference paper are well described and that no human error interferes with the results. A common human bias interferes with the selection of the depths of the core plugs. While the oil company may request plugs at exact intervals, it is not rare that the core laboratory would select them at depths where they are more easily extracted. [2]

19.2 COMBINING DATA FROM DIFFERENT SOURCES

Considering measurements from different sources with their associated uncertainties is an elegant solution to the endless arguments about "which measurement is right?" Once intervals of confidence are introduced, it can be seen (Fig. 19.1) that measurements of the same parameter may agree within their uncertainty range. The true value of the concerned formation parameter belongs to the overlap of the different intervals.

Conversely, forcing one measurement to another one, judged a priori of better quality than the others is often linked to a loss of information. If the reference measurement is so correct, why other measurements have been taken? Through the matching process, these additional measurements have become redundant.

1. It is only a fair idea as reproducibility implies that identical processes are used. In the experiment described in the reference, four different methods are used. A logging analogy would be to compare resistivities obtained with an induction device and with a laterolog device.
2. The spots where the rocks are easily breakable would be avoided if at all possible.

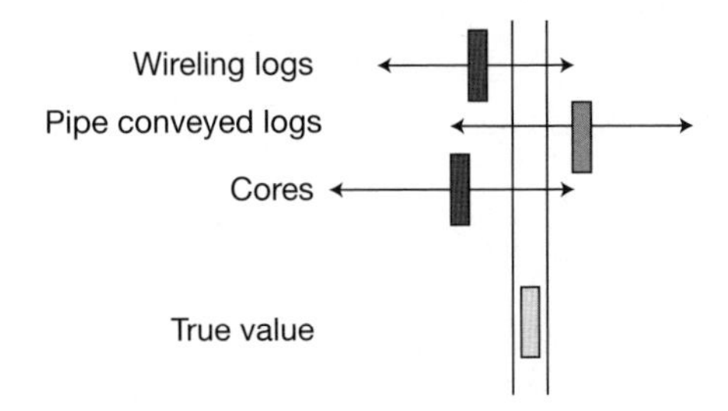

Figure 19.1

Three measurements (wireline, LWD and core-derived) do overlap once their corresponding uncertainties are accounted for. The true value, unknowable, is likely to belong to the overlap of the different intervals.

19.3 SUMMARY

- Core measurements are not more reliable than other measurements.
- Performing measurements through different contractors highlights large differences.
- It is easy to conclude that all measurements need to be integrated.
- Matching one measurement to another one is discouraged. Each measurement needs to be weighted by its uncertainty.

REFERENCES

[1] Neuman, C., "Logging measurement of residual oil, Rangely field," paper SPE 8844.
[2] Theys, P., *Log data acquisition and quality control*, chapter 27, Éditions Technip, 1999.

20

Conclusions and recommendations

If it takes thirty years to grow trees, we have to plant them today!
Theodore Roosevelt (1858-1919)

Formation evaluation has two ingredients, measurements and models. Petrophysical models have been given considerable attention for years. An abundance of analysis tools offer limitless model options and juggle with minimization and other forms of optimization. Conversely, well measurements are taken for granted and are not given the detailed attention they deserve. Incorrect nomenclatures such as calling a measured value a "formation parameter" still mislead many log analysts.

In the early decades of logging, log analysts had an intimate knowledge of the tool physics, technology and the data flow chart between raw data and delivered data. They were using this information along the interpretation process. Consciously or unconsciously, they were integrating implicit uncertainties in the computations. Due to the complexity of the wells and the complication of the data products and logging tools, this era is past.

Many chapters of this book may be unsettling [1]. Handling measurements is not as easy as it looks and there is much work needed to improve the situation. In the past, the burden of accomplishing progress was believed to be on the data vendor [2]. In fact, more efficient and complete data processes also require the active involvement of the users. These improvements necessarily require time. It is therefore important to initiate these efforts as soon as possible. Professional societies such as the Society of Petrophycisists and Log Analysts (SPWLA) are the obvious forums to facilitate these undertakings along the following lines:

1. One of the most indirect compliment made on a course on log measurement, was: Since this course, I have not slept well.
2. When DLIS was designed, oil companies' participation was limited. A Special Interest Group on Data Improvements was created in the early 2000s. It rapidly lost momentum as data users did not attend any longer when they understood that the task under attack was… huge.

- Data vendors and users need to dissipate the ambiguity between real and measured values.
- Standardize, then standardize and finally standardize should be the continuous mindsets of the industry.
- The critical role of the field engineer needs to be re-emphasized. His position vis-à-vis his employer, the data vendor and to his customer, the oil & gas company, requires to be reinforced so that he is no longer subjected to pressure to modify data or to under-report acquisition incidents. The logger's oath, giving the opportunity to the field personnel to act ethically without pressure, needs to be introduced.
- Major investments are required to improve the corporate databases. This means more complication in the short term, but a better usability in the long term.
- Improved documentation must be provided by data vendors.
- The interpretation analysis tools need to make a more complete use of the input data, in particular by handling uncertainties.

These recommendations apply to any type of well measurement, logs from open-hole or cased-hole, drill-pipe- or wireline-conveyed, as well as core and other drilling-related measurements.

20.1 DIFFERENCE BETWEEN REAL AND MEASURED VALUES

20.1.1 No misnomer

No well measurement should ever be called a formation parameter. The term "measured" should be used generously and added whenever possible as a natural attribute.

20.1.2 Start with a higher profile for measurement specifications

Once the data user fully understands the vendors' specifications, then the confusion between measurements and real values is greatly reduced. A strong emphasis on specifications, possibly highlighted in the deliverables, is a good first step.

20.1.3 The path to errors and uncertainties

Even further, errors and uncertainties must be unveiled and discussed without negative connotation. Any important measurement should be delivered with an "error" or "uncertainty" channel.

Data vendors have an extensive knowledge on the errors and uncertainties of the measurements acquired by their logging tools. They have an ethical duty to share this knowledge with the data users.

20.2 STANDARDIZATION

20.2.1 Similar structure of deliverables

Users and vendors groups need to convene to achieve as much standardization as possible on deliverables. In every data set, there exists the null value, a bit of information that contains no proprietary secret. Yet, it is different from vendor to vendor (Table 20.1).

Table 20.1 Representation of null values by different vendors

Vendor	Schlumberger	Baker	Baker (integer)
Null value	– 999.25	– 9,999.000	– 3,2767

Further, it is very difficult if not impossible to understand why vendors' delivery varies by two orders of magnitude for what appears to be the same measurement. [3]

It is also very unnerving to the end user when the data and information on the print are clearly not from the same source as the delivered digital medium [4]. To solve this issue, the graphical file (what used to be called print) must be created directly from or simultaneous with the digital file. There should be complete coherence between the two.

Chapter 14 provides a complete list of deliverables. It is the ground foundation for future delivery by all data vendors.

20.2.2 No danger of commoditization

The lack of standards in the oil industry is often explained by vendors as the result of the proprietary and competitive advantages supplied by a given data product. The validity of this statement is lacking when, in many industries, there is a high level of standardization, even though the products are highly differentiated. Most cars have wheels, clutches, cylinders, etc., following harsh regulatory constraints but a Ferrari can fetch 100 times the price of a Tata Nano. [4] The early setting of standards for the Compact Disk by two fierce competitors, Sony and Philips, has largely contributed to the success of this medium.

20.3 THE LOGGER'S OATH

The importance of the field engineers and of their direct supervisors has been emphasized in Chapter 17. These persons are under multiple sources of pressure that could easily result in conflicts of interest and interfere with their ability to properly report anomalous incidents.

3. In a specific example, 2 to 170 indexed channels are provided to the data user.
4. 2,200 US$ in 2010.

Field engineers have a privileged position to capture these details and their duty should be to pass on this information to the data users.

Field engineers and managers should swear to practice logging ethically, independently of their company affiliation.

20.4 IMPROVED DATABASES

There are large disparities in corporate databases that are well demonstrated when two companies merge. Standardization regarding databases is also on the agenda.

Conflicts between database managers and data users are very serious and intense. The first group wants the database to be simple and small and considers it as a cost. The second group would like the database to be broad in scope and content and see it as an asset. The long term vision should prevail. The management and capture of metadata is an important consideration during the design of the underlying data schemas. [5] It is repeated here that metadata is all the auxiliary information that enables a meaningful use of the data [1]. Other industries handling massive quantities of data (banks, insurance companies, hospitals) have well experienced that quality of data is mostly linked to the quality of metadata. [6] When there is insufficient metadata, the information sources cannot be understood any longer. In the long term, the acquisition data cannot be used and is eventually deleted.

20.5 DOCUMENTATION

Data products are complicated and sophisticated. But they do not come with much documentation. Data sets are commercialized even though basic information on the deliverables is poor. [7]

Essential documents, such as Quality Control reference manuals [2] & [3], environmental corrections charts and calibration guides need to be available to the data users as soon as new measurements are produced commercially by the logging tools.

---

5. All metadata can be lost when stored in some database and written out again [5].
6. A basic example of the need for metadata is linked to units. Units (such as m, ft, psi, kPa) are metadata. Some oil companies have made tragic and expensive mistakes because they were unsure about which unit was used during the acquisition of data. In the absence of units, they had to guess one (e.g., psi instead of kilopascals) – a bad example of a lottery with low winning odds – and hence took bad decisions.
7. The mnemonics of recently introduced tools are difficult to find on the vendors' websites. A basic contractual requirement should be that a measurement cannot be invoiced unless the deliverables are explicitly described.

20.6 THE REQUIRED EVOLUTION OF INTERPRETATION SOFTWARE

The current interpretation process is flawed as input data is assumed to be perfect and extremely precise. It is neither. Interpretation software should allow for inputting measurement uncertainties and propagate them through the evaluation process. These analysis tools should definitely give a higher profile to the acquisition metadata.

20.7 BEGINNING OF THE JOURNEY

This chapter concludes the road map of the quest for quality data. The map is only a tool, and the road to collect and use data is yet to be taken. The data vendors and users are invited to start this challenging journey.

REFERENCES

[1] Erpanet, Getting what you want, knowing what you have and keeping what you need. Metadata in digital preservation.

[2] Schlumberger, *Wireline Log quality Reference manual*, 1992.

[3] Schlumberger, *Anadrill Log quality Reference manual*, 1994.

[4] Storey, M., personal communication within the Oil Data Quality work group on Linkedin.

[5] Terry, R., personal communication within the Oil Data Quality work group on Linkedin.

Appendix

Appendix 1: Quantifying the level of proficiency in data quality

Oil & gas companies and other organizations using data can use the following checklist to rate the data vendors. The more points, the more proficient is the vendor.

1. SPECIFICATIONS

1.1 Measurements have metrological specifications (i.e. relevant to the measurement, not to the operating conditions of the tool).

From 1 to 7. 1: None. 7: for all measurements.

1.2 Are the sources (documentation) of specifications quantitatively documented for all measurements?

From 1 to 7. 1: None. 7: for all measurements.

1.3 Are the conditions of applications of the specifications carefully detailed?

From 1 to 7. 1: None. 7: for all measurements.

1.3 Are the sensitivities to the well conditions listed for the measurements?

From 1 to 7. 1: None. 7: for all measurements.

1.4 Is there a document listing all environmental corrections?

From 1 to 7. 1: None. 7: for all measurements and all conditions.

1.5 Are chartbooks available?

From 1 to 7. 1: None. 7: for all measurements.

1.6 Date of publication:

1: More than 10 years. 7: Less than two years.

1.7 Are flowcharts detailing the acquisition data processing chain available?

From 1 to 7. 1: None. 7: for all measurements.

2. DATA MANAGEMENT

2.1 Are all channels and parameters enabling traceability delivered to customers? From 1 to 7. 1: None. 7: for all measurements.

2.2 Are channels that have no information value removed from records delivered to customers?
From 1 to 7. 1: None. 7: for all measurements.

2.3 Are QC channels delivered to customers?
From 1 to 7. 1: None. 7: for all measurements.

3. EQUIPMENT

3.1 Is the auxiliary equipment dictated by the tool designers always available in field locations?
From 1 to 7. 1: None. 7: for all measurements.

3.2 Is the calibration equipment dictated by the tool designers always available in field locations?
From 1 to 7. 1: None. 7: for all measurements.

3.3 Is the calibration equipment always in top shape?
From 1 to 7. 1: None. 7: for all measurements.

4. PEOPLE

4.1 What is the definition of quality used by the logging company?
Meeting specified requirements: 7
Meeting client expectations: 2
Exceeding client expectations: 1

4.2 Is there a procedure preventing employees from tampering with log data?
Yes: 7; No: 1

4.3 Is this procedure well known by employees and adequately deployed?
Yes: 7; No: 1

4.4 Is there a centralized method to monitor the quality of the deliverables?
Yes: 7; No: 1

4.5 What is the most recent average rating available to management?
Between 75 and 85: 7; between 85 and 90: 6; between 90 and 95: 5; above 95: 2;
Below 75: 5

4.6 How often are products reviewed by an auditing team?
Never: 1; Every year: 7

Appendix 2: Deliverables

The following table (Table A2.1) lists the components of the graphical display of a log.

Notes:

1 – The items between brackets are suitable for pressure tests and LWD logs.

2 – The main log is always delivered. Generally, the equivalent digital file is loaded for formation evaluation. It is often the only component that is carefully analyzed by petrophysicists.

Tableau A2.1 List of graphical log components

Order	Component
1	Header
2	Remarks section
3	Log chronology
4	Depth information Box
5	Tool sketch
6	Well plot
7	Survey listing
8	Parameter listing
9	Parameter change box
10	Calibration information box
11	**MAIN LOG**
12	Repeat passes
(13)	Time based logs
(14)	Time –depth plot
15	QC logs/plots/flags
16	Tail
17	LQC stamp

Appendix 3: Lexicon

Integrity Policy: A high-level statement that data should not be nudged, modified or "cosmetized" by anybody, even under management or oil company pressure.

Operating specifications: Conditions under which a logging tool will function. Examples are temperature, pressure and hole size.

Metrological specifications: Quantification of the difference between the measurement and the real value. Also defines the volume of investigation of the logging tool. Example: accuracy, precision, depth of investigation, vertical resolution. Should also clearly define the conditions where the specifications are met.

Calibration guide: Any un-calibrated measurement is only qualitative. The guide describes the calibrators and calibration processes and frequency.

Correction charts: Charts describing the effects of the well environment and how these effects are compensated. Previous publications have often been misnamed: interpretation charts.

Reference manual; User-friendly and concise document gathering the checks to be performed on data to validate it.

Traceability flowchart: A chart showing the different steps in signal processing for a given measurement. It also includes the parameters and options that impact the outputs of the processing. These flowcharts do not need to detail proprietary software.

Mnemonics dictionary: A dictionary describing all the data objects delivered on a digital customer set.

Quality stamp: A checklist attached to the data product indicating that it has been checked by the person producing it and possibly by a supervisor.

Quality audits: Document reporting audits performed on data products and verifying that the quality stamp reflects a true assessment of the product.

Appendix 4: Metrological definitions

The number between brackets refers to the reference document. The other numbers relate to the article in the said reference. A definition may originate from different reference documents.

Accuracy of the measurement (ISO 10012 3.6, VIM 3.05)

The closeness of the agreement between the result of a measurement and the (conventional) true value of the measurand.

Notes: (1) Accuracy is a qualitative concept.
(2) The use of the term "precision" for "accuracy" should be avoided.

Errors (source Wikipedia)

- Observational error is the difference between the measured value and the true value. It is not a "mistake." Variability is an inherent part of things being measured.
- Systematic errors (or bias) and random errors are the main types.

Influence quantity (ISO 10012, 3.5)

A quantity which is not the subject of the measurement but which influences the value of the measurand or the indication of the measuring instrument.

Examples: ambient temperature; frequency of an alternating measured voltage.

Measurand (ISO 10012, 3.4, VIM 2.09)

A quantity subjected to measurement.

Precision (ISO 4259, 3.14)

The closeness of agreement between the results obtained by applying the experimental procedure several times on identical materials and under prescribed conditions. The smaller the random part of the experimental error, the more precise the procedure.

Reference conditions (ISO 10012, 3.13)

Conditions of use for a measuring instrument prescribed for performance testing, or to ensure valid intercomparison of results of measurements.

Note: The reference conditions generally specify "reference values" or "reference ranges" for the influence quantities affecting the measuring instrument.

Repeatability (ISO 4259, 3.17)

(a) Qualitatively: the closeness of agreement between independent results obtained in the normal and correct operation of the same method on identical test material, in a short interval of time, and under the same test conditions (same operator, same apparatus, same laboratory).

The representative parameters of the dispersion of the population which may be associated with the results are qualified by the term "repeatability," for example repeatability standard deviation, repeatability variance.

(b) Quantitatively: the value equal to or below which the absolute difference between two single test results obtained in the above conditions may be expected to lie with a probability of 95%.

Reproducibility (ISO 4259, 3.19)

(a) Qualitatively: the closeness of agreement between individual results obtained in the normal and correct operation of the same method on identical test material, but under different test conditions (different operators, different apparatus, different laboratories).

The representative parameters of the dispersion of the population which may be associated with the results are qualified by the term "reproducibility," for example reproducibility standard deviation, reproducibility variance.

(b) Quantitatively: the value equal to or below which the absolute difference between two single test results on identical material obtained by operators in different laboratories, using the standardized test method, may be expected to lie with a probability of 95%.

Specified measuring range (ISO 10012, 3.12)

The set of values for a measurand for which the error of a measuring instrument is intended to lie within specified limits.

True value (ISO 4259, 3.24)

For practical purposes, the value towards which the average of single results obtained by n laboratories tends, as n tends towards infinity; consequently, such a true value is associated with the particular method of test.

Note: a different and idealized definition is given in ISO 3534, Statistics – Vocabulary and symbols.

Other definition (VIM, 1.18)

The value of a measurand that is completely defined.

Notes: (1) This is the result that would be obtained by a perfect measurement.
(2) True value is an idealized concept.

Uncertainty of measurement (ISO 10012, 3.7)

Result of the evaluation aimed at characterizing the range within which the true value of a measurand is estimated to lie, generally with a given likelihood.

Sources

[1] International Organization for Standardization, *ISO 4259: Petroleum products - Determination and application of precision data in relation to methods of test*, Genève, 1992.

[2] International Organization for Standardization, *ISO 10012: Quality assurance requirements for measuring equipment*, Genève, 1992.

[3] International Organization for Standardization, *Guide to the expression of uncertainty in measurement (ISO/TAG4/WG3)*, Genève, 1995.

[4] International Organization for Standardization, *VIM 1984, International vocabulary of metrology, basic and general terms in metrology*, BIPM/CEI/ISO/OIML, Genève, 1984.

Appendix 5: Log Quality Control checklist

The following table contains an example of detailed checklist that can be used to audit graphical deliverables. The list is provided by courtesy of Schlumberger. A similar list can be built to perform a detailed analysis of any type of log. It is recommended that the same list is used by the data vendor and the data user.

Category	Descrition of the problem	Weight
Header	Minor omission, spelling errors	1
	Wrong template	5
	Major error: field, client name, witness, log name	2
Job event summary	Missing, incomplete, lack of details	5
Remarks	Anomalies in data not remarked	2
	Relevant information not mentioned, (e.g., source of maximum temperature not mentioned)	2
	Misleading remarks, spelling mistakes	1
Total depth, first and readings, casing shoe	Labels omitted	1
Tool sketch	Missing	5
	Errors	2
	Missing serial numbers	2
	Disagreement with parameter summary or special inserts	5
Well sketch	Missing	5
	Errors (only drillers' depth should be used)	2
Curves	Labels are missing	1
	Curves are missing (e.g., $\Delta\rho$)	5
	Scales, coding, areas are incorrect	2
Formats	Wrong logarithmic scales	5
	Wrong vertical scale	5
	Wrong pips (integration)	2
Parameter summary	Wrong parameters	5
	Parameter listing incomplete	5
	Parameter listing set incorret: too few or too many parameters	2
	Missing separators/spelling mistakes in separators	2
Parameter selection	Incorrect selection	5
	Important parameters set to default values	2
Print	Print not aligned or not folded properly	2
	Wrong order of log components	2
	Print too dark or too light, bleeding colors	2
Depth	Missing depth summary listing	5
	Wrong calibrators or cable serial number	2
	Unrealistic cable stretch	2

Category	Descrition of the problem	Weight
	Log off-depth as compared to reference log (<1 ft)	2
	Log off-depth as compared to reference log (>1 ft)	5
Tool positioning	Incorrect selection of standoff	5
	Incorrect positioning (e.g., poorly centralized)	5
Logging speed	Incorrect choice (e.g., for high-resolution logs)	5
Digital record	Verification listing not available	5
	File numbers do not match print	2
	Poor or incomplete tape/CD/DVD labels	1
	Missing channels	5
Repeat section	Missing or too short (minimum 200 ft or 70 m)	5
	Log does not repeat within specification	2
	Anomalies are not annotated or remarked	5
Response in known conditions	Proper labeling (e.g., checks in casing)	1
	Unsatisfactory check (e.g., in salt or anhydrite)	5
Failure, log anomalies	Found in real time, repeated, commented	1
	Not repeated but remarked	2
	Not repeated/not commented.	5
QC curves	Missing	5
	Indicate anomalies but not remarked	2
	Show anomalies and remarked/explained properly	1
Calibrations	Missing, incomplete, lack of details	5
	Out of tolerance, expired	5
Checks (before/after)	Out of tolerance, incorrect dates	2

Appendix 6: Advanced computation of uncertainties for the density log

The uncertainties given by the data vendors have been reviewed in Chapter 13. They come in addition to the uncertainties computed in this appendix. The density measurement has no correction chart attached to it. So, there is no propagated uncertainty from these charts.

The following contributions need to be accounted for:

- The uncertainty due to the hole shape and size.
- The uncertainty linked to the mud or mud cake compensation scheme (often called $\Delta\rho$).
- The uncertainty due to the hole rugosity.

Density uncertainty due to the hole diameter

$$slope_{CALI} = [1/(2*s)]\,[(CALI_n - CALI_{n-1}) + (CALI_{n+1} - CALI_n)]$$

$$\text{If } CALI \leq 9,\ \sigma_{CALI} = 0$$

$$\text{If } 9 < CALI \leq 16,\ \sigma_{CALI} = 0.002 * (CALI - 9)$$

$$\text{If } CALI > 16 \text{ and } slope_{CALI} < 0.1.\ \sigma_{CALI} = \sqrt{0.1}$$

$$\text{If } CALI > 16 \text{ and } slope_{CALI} > 0.1.\ \sigma_{CALI} = 0.002 * (CALI - 9)$$

Uncertainty originating from $\Delta\rho$

$$\text{If } \Delta\rho \geq 0,\ \sigma_{\Delta\rho} = 6 * \Delta\rho^2$$

$$\text{If } \Delta\rho < 0, \text{ and } w_{mud} \leq 12,\ \sigma_{\Delta\rho} = 6 * \Delta\rho^2 + |\Delta\rho|$$

Uncertainty due to the hole rugosity

$$H_r = (1/4 * s^2) *$$

$$[(CALI_{n-3} - 2 * CALI_{n-2} + CALI_{n-1}) + (CALI_{n-2} - 2 * CALI_{n-1} + CALI_n) +$$

$$(CALI_{n-1} - 2 * CALI_n + CALI_{n+1}) + (CALI_n - 2 * CALI_{n+1} + CALI_{n+2}) +$$

$$(CALI_{n+1} - 2 * CALI_{n+2} + CALI_{n+3})]$$

$$\sigma_{hole\ rugosity} = 0.1 * H_r^2$$

Combining uncertainties

$$\sigma_\rho = \sqrt{\sigma_{CALI}^2 + \sigma_{\Delta\rho}^2 + \sigma_{hole\ rugosity}^2}$$

Numerical application

Data has been acquired from 2,392.67 m up to 2,364.78 m. Bit size is 12.25 in. The hole is not smooth and the density correction shows some activity. The algorithms shown above are used and a "premium" uncertainty is calculated sample by sample. The components are σ_{cali}, $\sigma_{\Delta\rho}$ and σ_{hr}. The contribution of each uncertainty is shown in the last columns.

In the first intervals, the caliper uncertainty has the largest contribution. The main contributors are shaded in grey.

Comments

Once the algorithms are tested and frozen, it is very easy to quantify uncertainties. Only input data is needed. Uncertainty channels are welcome additions to the acquisition deliverables.

Tableau A6.1 Input data and uncertainty computation. The last columns represent the percentage of the uncertainty contributors.

Depth	Caliper	Density	$\Delta\rho$	σ_{cali}	$\sigma_{\Delta\rho}$	H_r	σ_{Hr}	σ_ρ	%cali	%$\Delta\rho$	%Rugo
2,364.78	13.032	2.194	0.037								
2,364.93	12.293	2.186	0.023								
2,365.08	12.739	2.199	0.012	0.0075	0.0009						
2,365.24	13.062	2.224	0.009	0.0081	0.0004	0.0469	0.0002	0.0081	99.6	0.3	0.1
2,365.39	13.201	2.211	0.004	0.0084	0.0001	-0.0611	0.0004	0.0084	99.8	0.0	0.2
2,365.54	13.397	2.195	0.006	0.0088	0.0002	-0.0442	0.0002	0.0088	99.9	0.1	0.0
2,365.69	13.409	2.179	0.009	0.0088	0.0004	-0.0005	0.0000	0.0088	99.7	0.3	0.0
2,365.85	12.878	2.182	0.009	0.0078	0.0005	-0.0060	0.0000	0.0078	99.6	0.4	0.0
2,366.00	12.493	2.190	0.008	0.0070	0.0004	-0.0019	0.0000	0.0070	99.7	0.3	0.0
2,366.15	12.624	2.196	0.013	0.0072	0.0010	0.0325	0.0001	0.0073	98.1	1.9	0.0
2,391.75	12.724	2.207	0.078	0.0074	0.0362	0.0240	0.0001	0.0370	4.1	95.9	0.0
2,391.91	12.705	2.209	0.083	0.0074	0.0410	0.0010	0.0000	0.0417	3.2	96.8	0.0
2,392.06	12.693	2.206	0.083	0.0074	0.0416	-0.0024	0.0000	0.0423	3.1	96.9	0.0
2,392.21	12.693	2.208	0.084	0.0074	0.0427	-0.0041	0.0000	0.0434	2.9	97.1	0.0
2,392.36	12.839	2.212	0.082	0.0077	0.0404	-0.8003	0.0640	0.0761	1.0	28.2	70.8
2,392.52	12.901	2.208	0.076	0.0078	0.0348						
2,392.67	12.816	2.204	0.069	0.0076	0.0282						

Appendix 7: Quest for mnemonics

The quest for information starts with an understanding of the data objects. The first step is to recognize the name of the data object and link it to its definition and description. The oil industry uses the term "mnemonics" for these names. This appendix gives some clues how to find mnemonics for the major data vendors. The information in this appendix is bound to be modified at very short notice. The approach with the highest probability of success is to search internet with the vendor's name added to the term "mnemonics."

The worst drawback of the current mnemonics format is that the user needs to know in advance what is the "mnemonic" of a data object, then search for it. This format is of little utility to a person trying to find out if his data set is complete, that if, if any data object is missing.

Baker Inteq

http://www.bakerhughesdirect.com/cgi/inteq/INTEQ/ServiceLib/DisplayInfo/serviceMnemonics.jsh?index=F

Halliburton

www.halliburton.com/ps/default.aspx?pageid=336

Schlumberger

http://www.apps.slb.com/cmd/

The challenge with this vendor is to find the mnemonic name, which most of the time does not coincide with the tool name or the commercial name (Table A7.1).

Table A7.1. Example of complication with the names of measurements

Measurement	Commercial name	Technical name	Mnemonics name
3-D Induction	Rt Scanner	?	ZAIT-BA
Acoustic	Sonic Scanner	MSIP	MAST, MAPC, MAPC-A, MAPC-B
Nuclear magnetic resonance	MRScanner	MRX	MRX

Weatherford

http://www.weatherford.com/weatherford/groups/web/documents/weatherfordcorp/WFT100552.pdf

Imprimé en France en janvier 2011 par EMD S.A.S.
53110 Lassay-les-Châteaux
N° d'imprimeur : 24437 - Dépôt légal : janvier 2011